Digital Control Systems

Lecture Notes 2021

Thomas Paul Weldon

CONTENTS

PREFACE

The updated collection of lecture notes in this book is based on over 20 years of teaching graduate and undergraduate courses at the University of North Carolina at Charlotte. As lecture notes, this book is not intended to be a substitute for the many excellent textbooks in this field. Instead, this book is intended as a supplement to other course materials and as a workbook for students taking notes during corresponding lectures. In addition, practicing engineers may find this book useful for quick review of the topic.

Thomas Paul Weldon
Charlotte, NC
July 12, 2021

1 INTRODUCTION AND REVIEW

The lecture notes in this chapter provide a brief introduction to digital control systems and a review of fundamentals of analog control systems.

Overview

- Digital control systems
  - Some portion of control loop is digital
  - Loop typically has both digital and analog signals/systems
  - Typically include ADC (analog-to-digital converter) and DAC (digital-to-analog converter)
  - Incorporate digital signal processing
    - Digital loop filter to control gain/bandwidth/overshoot/etc.
    - Hardware or software signal processing
    - Can be adaptive
- Advantages
  - Replace/improve legacy analog control systems
  - Advantages in adaptive to changing conditions
  - Reconfigurable/reprogrammable for upgrades

Outline

- Outline of topics
  - Brief review of analog control systems basics and Laplace transform
  - Review of digital signal processing and z-transform
  - Starred transform analysis of mixed analog/digital systems
  - Open-loop digital control systems
  - Closed-loop digital control systems and stability
  - Digital lag, lead, and PID compensators
  - Applications to digital impedance and phase-locked loops
  - Digital state-variable models, methods, and analysis
  - Digital observer and controller design
  - System identification basics

Analog Control System Basics/Review

Brief Review of Analog Control Systems

- Analog control systems
  - o Basic idea:
    - – Use control/feedback to improve some device (motor)
  - o Examples:
    - – Improve bandwidth or gain of an amplifier
    - – Control speed or position of a motor
- Typical analog control system
  - o Has device that is to be controlled/enhanced
  - o Has input reference signal (voltage, position, phase)
  - o Uses some sort of feedback (perhaps from a sensor)
  - o Measures error (reference compared to feedback)
  - o Based on error, make some correction to device input

Unit Circle Review

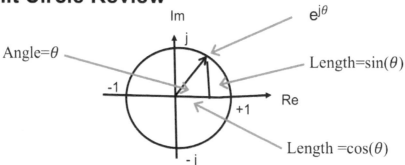

- $e^{j\theta} = \cos(\theta) + j \sin(\theta) = \alpha + j\beta$
- Can be viewed as the point in complex plane with coordinates ($\cos\theta$, $\sin\theta$)
- Can also be viewed as a vector, as shown above
- $|e^{j\theta}| = ?$

Unit Circle

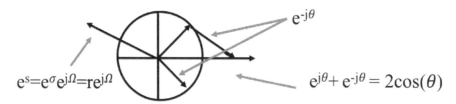

- As shown above, vector viewpoint is useful when adding
- Phasor notation can also be useful in polynomials
- Also e^s or e^{st} can be shown as above, where $s = \sigma + j\Omega$
- Consider finding square root of j

$(\gamma \angle \theta)^2 \;=\; \gamma^2 \angle 2\theta \;=\; j \;=\; 1\angle 90°$

or $(\gamma\, e^{j\theta})^2 \;=\; \gamma^2\, e^{j2\theta} \;=\; j \;=\; 1\, e^{j\pi/2}$ (assuming γ is positive real)

- So, there are 2 roots to 2nd order polynomial

$1\angle 45°$ AND $1\angle 225°$

Analog Continuous-Time Feedback Control System

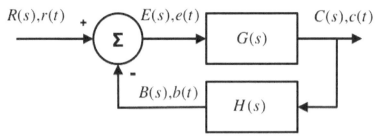

- Input R(s), reference signal
- Output C(s), controlled signal
- Feedback B(s), feedback signal (back)
- Error E(s), error signal (ideally zero if C(s)=R(s))
- G(s) is forward path transfer function (or plant function)
- H(s) is feedback transfer function
- $G_{OL}(s)=G(s)H(s)$ is the open-loop gain

Continuous-Time Feedback Control System

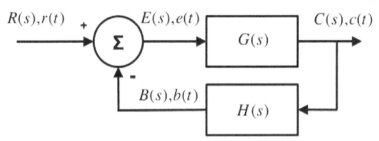

- Relations for error E(s) and output C(s):

$$E(s) = R(s) - B(s) \qquad B(s) = C(s)H(s)$$

$$C(s) = E(s)G(s) = \{R(s) - B(s)\}G(s)$$

$$= \{R(s) - C(s)H(s)\}G(s)$$

- Solving for closed-loop gain (closed-loop transfer function) $G_{CL}(s)$:

$$G_{CL}(s) = \frac{C(s)}{R(s)} = \frac{G(s)}{1 + G(s)H(s)}$$

Example Cont.-Time Feedback Control System

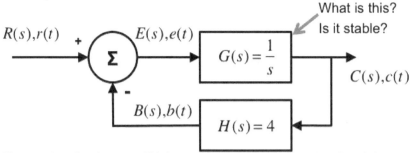

What is this?
Is it stable?

- Example: find error E(s) and closed-loop gain $G_{CL}(s)$:

$$G_{CL}(s) = \frac{C(s)}{R(s)} = \frac{G(s)}{1+G(s)H(s)} = \frac{1/s}{1+4/s} = \frac{1}{s+4} \quad \leftarrow \text{Stable lowpass}$$

$e^{-4t} u(t)$

$$\frac{E(s)}{R(s)} = \frac{R(s)-B(s)}{R(s)} = 1 - \frac{C(s)H(s)}{R(s)} = 1 - \frac{C(s)}{R(s)}H(s)$$

dc error=0

$$= 1 - \frac{G(s)H(s)}{1+G(s)H(s)} = \frac{1}{1+G(s)H(s)} = \frac{1}{1+4/s} = \frac{s}{s+4}$$

Bode Plots

$$\frac{V_{out}(s)}{V_{in}(s)} = \frac{R_2 + 1/sC}{R_1 + R_2 + 1/sC} = \frac{1+sR_2C}{1+s(R_1+R_2)C} = \frac{1+s/100}{1+s/10} \quad for \quad \begin{cases} R_1 = 90 \\ R_2 = 10 \\ C = 1/1000 \end{cases}$$

- Bode magnitude plots
  - Slope decreases 20dB/decade for each <u>pole</u> frequency
  - Slope increases 20dB/decade for each <u>zero</u> frequency
- Bode Phase plots
  - Phase (degrees) decreases 45° at at each <u>pole</u> frequency
  - Phase (degrees) increases 45° at at each <u>zero</u> frequency
  - Interpolated at ±45°/decade for ±one decade either side of pole or zero
- For example of simple RC circuit above:

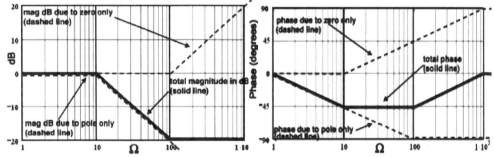

Example: Amplifier Feedback

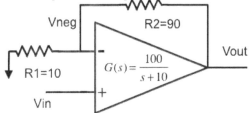

- Recall: no current into op-amp terminals, so
  - Vneg/R1 = Vout/(R1+R2)
  - Vout(s) = (Vin-Vneg(s)) G(s)

$$V_{out}/(R_1+R_2) = V_{neg}/R_1 = (V_{in}-V_{out}/G(s))/R_1$$

$$\frac{V_{out}(s)}{V_{in}(s)} = \frac{G(s)}{1+G(s)R_1/(R_1+R_2)} = \frac{G(s)}{1+G(s)H(s)}$$

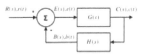

$$G_{CL}(s) = \frac{V_{out}(s)}{V_{in}(s)} = \frac{100/(s+10)}{1+1000/(s+10)/100} = \frac{100}{s+20} = 5@s=0$$

Example: Amplifier Feedback Bode Magnitude Plot

$$G_{CL}(s) = \frac{V_{out}(s)}{V_{in}(s)} = \frac{G(s)}{1+G(s)H(s)} = \frac{100}{s+20}$$

$$G(s) = \frac{100}{s+10} \quad G(s)H(s) = \frac{10}{s+10}$$

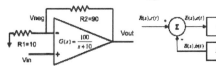

- Plot |G(s)|, |G(s)H(s)|, and closed-loop gain |G$_{CL}$(s)| in dB
  - Recall: Bode magnitude breaks 20dB/decade at 3dB point

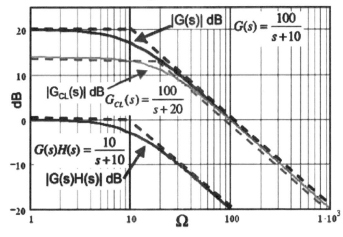

Bode plot analysis of previous example

Example: Amplifier Feedback Bode Phase Plot

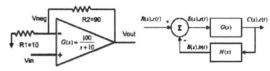

$$G_{CL}(s) = \frac{V_{out}(s)}{V_{in}(s)} = \frac{G(s)}{1+G(s)H(s)} = \frac{100}{s+20}$$

$$G(s) = \frac{100}{s+10} \quad G(s)H(s) = \frac{10}{s+10}$$

- Plot $\angle G(s)$, $\angle G(s)H(s)$, and closed-loop phase $\angle G_{CL}(s)$
  - Recall: Phase (degrees) is 45° at 3dB point, 45°/decade

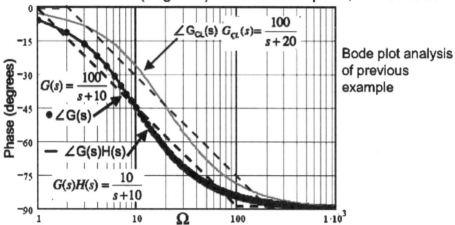

Bode plot analysis of previous example

Overview of Digital Control Systems

Simple Digital Control System: Add ADC and DAC

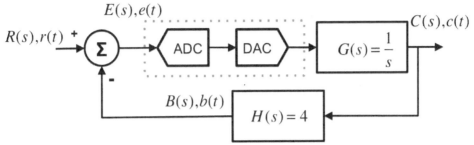

- Suppose everything the same, except:
  - o Add an ADC and DAC as shown above
  - o Suppose excellent ADC/DAC: 100 megasample/s, 20-bit
  - o Is the response the same???
  - o What about Nyquist sampling/aiasing?
  - o Fastest digital frequency: on-off…= 1,0,1,0,1…
- Loop now has BOTH analog and digital signals and circuits
  - o Analysis math must handle analog and digital signals

Digital Sinusoid Aliasing in Time

- Consider an analog sinusoid with extra 2π between samples
- Samples are exactly the same
- So, two different analog signals give rise to same samples
- To avoid aliasing, must have $1/T_S = f_S > 2 f_0 = 2/T_0$ (Nyquist)

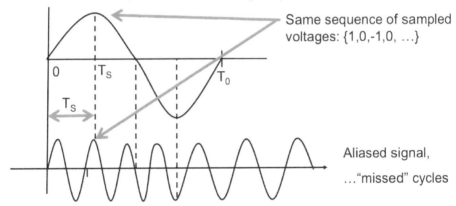

Same sequence of sampled voltages: {1,0,-1,0, …}

Aliased signal, …"missed" cycles

...In the ADC and DAC example

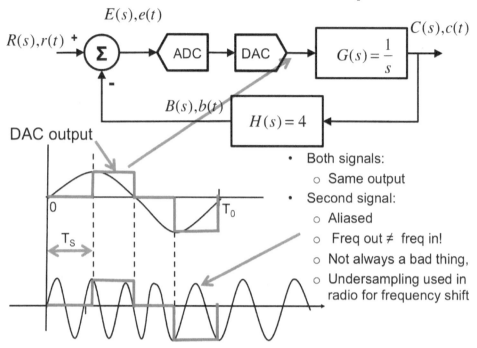

- Both signals:
  - Same output
- Second signal:
  - Aliased
  - Freq out ≠ freq in!
  - Not always a bad thing,
  - Undersampling used in radio for frequency shift

Motivation for Study of Digital Control Systems

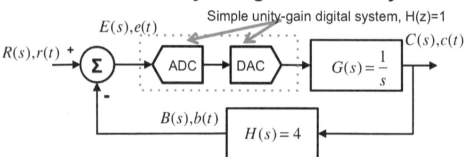

- This simple example illustrates motivations for digital control
  - The effect of ADC+DAC is not as simple as it may appear
  - Speed limit: Nyquist limit (frequency < sample rate/2)
  - Aliasing beyond Nyquist: output freq. ≠ input frequency!
  - Filter responses: frequency warping or aliasing
  - Some signals in loop are analog, some are digital
  - Must blend z-transform and Laplace transform in analysis
  - Stability analysis must include all issues above

2 LAPLACE TRANSFORM

The lecture notes in this chapter provide a brief review of Laplace transform methods, and an overview of complex analysis for inverse Laplace transforms.

Laplace Transform

Review of Laplace Transform

- Pierre-Simon Laplace invented ~1825/1845
- Review of Laplace transform
  - o First: eigenfunction e^{st} has key role in differential equations
  - o So Laplace transform has huge role in many applications
  - o Laplace methods simplify solution of differential equations
  - o Later: z-transform serves similar role for digital systems
- We will primarily use one-sided Laplace transform
  - o But also mention two-sided Laplace transform
- We will review use of partial fraction expansion for inverse Laplace transform

One-Sided (Unilateral) Laplace Transform
- Laplace transform
  - Pierre-Simon Laplace invented ~1825/1845
- Define 1-sided (unilateral) Laplace transform:

$$X(s) = L\{x(t)\} = \int_{\varepsilon \to 0}^{\infty} x(t)e^{-st}\,dt = \int_{0^-}^{\infty} x(t)e^{-(\sigma+j\Omega)t}\,dt$$

$$x(t) = L^{-1}\{X(s)\} = \frac{1}{j2\pi} \int_{\sigma-j\infty}^{\sigma+j\infty} X(s)e^{st}\,ds$$

- The inverse transform is a contour integral in complex plane
- We will consider inverse Laplace a bit more deeply
- We will later see that z-transform is much the same
- **NOTE**: notation "$\sigma+j\Omega$" for Laplace, NOT "$\sigma+j\omega$" as in text

Example Laplace Transform and Inverse
- Example: let $x(t)= e^{-at}\, u(t)$
- 1-sided Laplace transform :

$$X(s) = \int_{\varepsilon \to 0}^{\infty} e^{-at}u(t)e^{-st}\,dt = \int_{0^-}^{\infty} e^{-(s+a)t}\,dt = \left. \frac{e^{-(s+a)t}}{s+a} \right|_{0^-}^{\infty}$$

ROC
(region of convergence)

$$= \frac{1}{s+a} \text{ for } \boxed{\text{Re}\{s\} > -\text{Re}\{a\}}$$

- The inverse transform:

$$x(t) = \frac{1}{j2\pi} \int_{\sigma_i-j\infty}^{\sigma_i+j\infty} \frac{1}{s+a}e^{st}\,ds = ???$$

Im(s)

ds

Re(s)

-a   0 σ_i

Line of
integration

- Note: σ must be in ROC region where X(s) is defined
- So, need "complex analysis" to do inverse transform
- Or, use lookup tables and partial fraction expansion

Table of Some 1-Sided Laplace Transforms

Time Function	Laplace Transform
$\delta(t)$	1
$u(t)$	$1/s;\ \mathrm{Re}\{s\}>0$
$tu(t)$	$1/s^2;\ \mathrm{Re}\{s\}>0$
$e^{-at}u(t)$	$\dfrac{1}{s+a};\ \mathrm{Re}\{s\}>-a$
$te^{-at}u(t)$	$\dfrac{1}{(s+a)^2};\ \mathrm{Re}\{s\}>-a$
$\cos(\Omega_0 t)u(t)$	$\dfrac{s}{s^2+\Omega_0^{\,2}};\ \mathrm{Re}\{s\}>0$
$\sin(\Omega_0 t)u(t)$	$\dfrac{\Omega_0}{s^2+\Omega_0^{\,2}};\ \mathrm{Re}\{s\}>0$

Two-Sided (Bilateral) Laplace Transform Definition

- Although we will not use it, it is important to note the two-sided Laplace transform
- More general: allows non-causal x(t)
- But, MUST include ROC (region of convergence)
- Note: book uses 1-sided Laplace
- Define the two-sided Laplace transform as:

$$X(s)=L\{x(t)\}=\int_{-\infty}^{\infty} x(t)e^{-st}\,dt = \int_{-\infty}^{\infty} x(t)e^{-(\sigma+j\Omega)t}\,dt \leftarrow$$

$$x(t)=L^{-1}\{X(s)\}=\frac{1}{j2\pi}\int_{\sigma-j\infty}^{\sigma+j\infty} X(s)e^{st}\,ds$$

- MUST also include ROC (region of convergence) for unique transform to distinguish left-sided x(t) from right-sided x(t), e.g., to distinguish $e^{-at}u(t)$ from $e^{at}u(-t)$

Review of Differential Equations

- Differential equations
    - Found to model many systems including control systems
- One popular solution method:
    - Method of judicious guessing
    - Guess e^{st} is the solution of differential equation
    - Solve for coefficients
- Example:

$$\frac{d^2}{dt^2}g(t)+k\frac{d}{dt}g(t)=0 \quad \Rightarrow \quad s^2 e^{st}+ks\,e^{st}=s(s+k)e^{st}=0$$

$$so:\ s(s+k)=0 \ \Rightarrow \ g(t)=Ae^0+Be^{-kt}$$

and use initial conditions to solve for A & B

Laplace leads to same coefficients relation: s(s+k)

$$\text{Recall: } L\left\{\frac{d^2}{dt^2}g(t)+k\frac{d}{dt}g(t)=0\right\} \Rightarrow (s^2+ks)G(s)=0$$

e^{st} Are Eigenfunctions of Differential Eqations

- Why does the "judicious guess" e^{st} work so well?
    - The "guess" solution e^{st} is an eigenfunction of derivatives
- Eigenfunctions are the special functions where output of some operation has same form as input of the operation
- Eigenvalues are the "scale factor" by which the eigenfunction is scaled (can be complex or real)
- To see this: $\dfrac{d}{dt}e^{st}=se^{st}$ $\quad so:\ s$ is eigenvalue, e^{st} is eigenfunction
- As noted earlier, Laplace transform yields similar results
- Also note, engineers prefer Z(s)=1/(sC) over Cdv/dt!

$$i(t)=C\frac{dv(t)}{dt} \quad \Rightarrow \quad Z(s)=\frac{V(s)}{I(s)}=\frac{1}{sC}$$

Laplace transform methods make the problems easier!!!

$$v(t)=L\frac{di(t)}{dt} \quad \Rightarrow \quad Z(s)=sL$$

Inverse Laplace transform

- Inverse Laplace transform
  - The inverse transform is a contour integral in complex plane
  - **<u>NOTE</u>**: notation "$\sigma+j\Omega$" for Laplace, NOT "$\sigma+j\omega$" as in text
  - Later, we will see that z-transform has similar role in digital systems
- Most common approach to inverse Laplace transform:
  - Use a look-up table
  - If the function is not in the table, do partial fraction expansion

Inverse Laplace Transform by Partial Fraction Expansion

- **Consider partial fraction expansion for first-order poles**

  1) Suppose X(s)=N(s)/D(s),

    - Where degree of N(s) < degree of D(s)

    - If not: First do long division until the degree of N(s) is less than the degree of D(s)

  2) Factor denominator

$$X(s) = \frac{N(s)}{D(s)} = \frac{N(s)}{(s-p_1)(s-p_2)\circ\circ\circ(s-p_i)} = \sum_{\alpha=1}^{i} \frac{c_\alpha}{s - p_\alpha}$$

  3. Multiply both sides by (s-p$_k$)

$$(s-p_k)\frac{N(s)}{D(s)} = \frac{(s-p_k)N(s)}{(s-p_1)(s-p_2)\circ\circ\circ(s-p_i)} = (s-p_k)\left\{\sum_{\alpha=1}^{i} \frac{c_\alpha}{s - p_\alpha}\right\}$$

Partial Fraction Expansion

4. At s=p$_k$ solve for c$_k$

$$(s-p_k)\frac{N(s)}{D(s)}\bigg|_{s=p_k} = c_k$$

6. Having solved for c$_k$ the partial fraction expansion is:

$$X(s) = \frac{N(s)}{D(s)} = \sum_{\alpha=1}^{i} \frac{c_\alpha}{s - p_\alpha}$$

7. Taking inverse Laplace transform

$$e^{-at}u(t) \Leftrightarrow \frac{1}{s+a}; \operatorname{Re}\{s\} > -a$$

Beware of sign differences

$$\text{So}: \quad x(t) = L^{-1}\{X(s)\} = L^{-1}\left\{\sum_{\alpha=1}^{i} \frac{c_\alpha}{s - p_\alpha}\right\} = \sum_{\alpha=1}^{i} c_\alpha e^{p_\alpha t} u(t)$$

Beware of sign differences

Partial Fraction Expansion Example

- Partial Fraction Expansion Example:

•First do long division, since numerator degree is NOT less than denominator

$$s^2+7s+12\overline{\smash{\big)}s^2+1}$$
$$s^2+7s+12$$
$$-7s-11$$

$$X(s)=\frac{s^2+1}{(s+3)(s+4)}=\frac{N(s)}{D(s)}\ long\,division \Rightarrow$$

$$so \quad X(s)=1+\frac{-7s-11}{(s+3)(s+4)}$$

$$=1+\frac{(s+3)\dfrac{-7s-11}{(s+3)(s+4)}\bigg|_{s=-3}}{s+3}+\frac{(s+4)\dfrac{-7s-11}{(s+3)(s+4)}\bigg|_{s=-4}}{s+4}$$

$$=1+\frac{\dfrac{-7s-11}{(s+4)}\bigg|_{s=-3}}{s+3}+\frac{\dfrac{-7s-11}{(s+3)}\bigg|_{s=-4}}{s+4}=1+\frac{10}{(s+3)}+\frac{-17}{(s+4)}$$

Partial Fraction Example, continued

- Rearrange partial fraction expansion results into forms where x[n] may be found using z-transform tables

- Use z-transform properties tables, as needed

- For preceding example:

$$X(s)=1+\frac{10}{(s+3)}+\frac{-17}{(s+4)}$$

$and:$

$$e^{-at}u(t)\Leftrightarrow\frac{1}{s+a}$$

$so:$

$$x(t)=\delta(t)+10e^{-3t}u(t)-17e^{-4t}u(t)$$

Laplace Poles and Stability

- The example H(s)=1/(s+4) has a single pole in the s-plane
- The system is BIBO stable (bounded-input bounded-output)
    - Because all poles are in left half-plane
    - Because $h(t)=e^{-4t} u(t)$ and $\int |h(t)| dt < \infty$
    - Because ROC includes $j\Omega$ axis
- If H(s)=1/(s-4), then $h(t)=e^{4t} u(t)$ diverges!

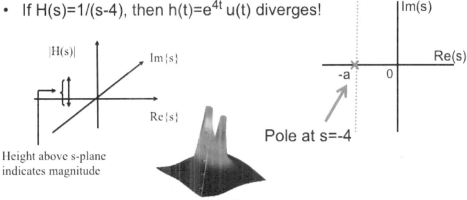

Height above s-plane
indicates magnitude

Pole at s=-4

Fourier Transform

- Fourier Transform

If g(t) causal, =1-sided, too

$$\mathcal{F}\{g(t)\} = G(\Omega) = \int_{-\infty}^{\infty} g(t)e^{-j\Omega t}\, dt = L_{2-sided}\{g(t)\}\big|_{s=j\Omega}$$

$$g(t) = \frac{1}{2\pi}\int_{-\infty}^{\infty} G(\Omega)e^{j\Omega t}\, d\Omega$$

- Fourier transform properties:

$$g_1(t)g_2(t) \Leftrightarrow \frac{1}{2\pi}G_1(\Omega)*G_2(\Omega)$$

$$g_1(t)*g_2(t) \Leftrightarrow G_1(\Omega)G_2(\Omega)$$

$$\Pi\left(\frac{t}{\tau}\right) = rect\left(\frac{t}{\tau}\right) \qquad \Delta\left(\frac{t}{\tau}\right)$$

$$\delta_{T_0}(t) = \sum \delta(t - nT_0)$$

- Examples

$$\Pi(t/\tau) = rect(t/\tau) \Leftrightarrow \tau\, sinc(\Omega\tau/2) = (1 - e^{-j\Omega\tau})e^{j\Omega\tau/2}/j\Omega$$

$$\Delta(t/\tau) \Leftrightarrow 0.5\tau\, sinc^2(\Omega\tau/4) \qquad and \qquad e^{j\Omega_0 t} \Leftrightarrow 2\pi\delta(\Omega - \Omega_0)$$

$$\delta_{T_0}(t) = \sum_{n=-\infty}^{\infty} \delta(t - nT_0) \Leftrightarrow (2\pi/T_0)\sum_{n=-\infty}^{\infty}\delta(\Omega - n2\pi/T_0) = \Omega_0\sum\delta(\Omega - n\Omega_0)$$

Beware: some texts use 1-sided impulse train

Laplace From Fourier

- Another note
  - Laplace can be created from Fourier transform
  - Jean-Baptiste Fourier invented ~1807 Fourier series g[11]
- Observe for causal x(t):

$$X(s) = L\{x(t)\} = \int\limits_{\varepsilon \to 0}^{\infty} x(t)e^{-st}\,dt = \int\limits_{0^-}^{\infty} x(t)e^{-(\sigma + j\Omega)t}\,dt$$

$$= \int\limits_{0^-}^{\infty} x(t)e^{-\sigma t}e^{-j\Omega t}\,dt = \mathcal{F}\{x(t)e^{-\sigma t}u(t)\}$$

g^9-11g

Example Use of Laplace for Control Loop

Example Feedback Control System

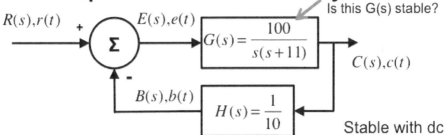

Is this G(s) stable?

$$C(s),c(t)$$

Stable with dc gain=10

- Find closed-loop gain $G_{CL}(s)$:

$$C(s) = (R(s) - C(s)H(s))G(s)$$

$$G_{CL}(s) = \frac{C(s)}{R(s)} = \frac{G(s)}{1+G(s)H(s)} = \frac{100/(s(s+11))}{1+10/(s(s+11))} = \frac{100}{s^2+11s+10} = \frac{100}{(s+10)(s+1)}$$

- Find error response E(s)/R(s):

dc error=0

$$\frac{E(s)}{R(s)} = \frac{R(s)-B(s)}{R(s)} = 1 - \frac{C(s)H(s)}{R(s)} = 1 - \frac{C(s)}{R(s)}H(s)$$

$$= 1 - \frac{G(s)H(s)}{1+G(s)H(s)} = \frac{1}{1+G(s)H(s)} = \frac{1}{1+10/(s(s+11))} = \frac{s(s+11)}{(s+10)(s+1)}$$

Example continued

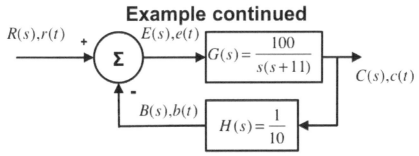

$$C(s),c(t)$$

- Find closed-loop impulse response $g_{CL}(t)$:

$$G_{CL}(s) = \frac{100}{(s+10)(s+1)} = \frac{100/9}{(s+1)} + \frac{-100/9}{(s+10)} \quad and: e^{-at}u(t) \Leftrightarrow \frac{1}{s+a}$$

$$so: g_{CL}(t) = L^{-1}\left\{\frac{100/9}{(s+1)} + \frac{-100/9}{(s+10)}\right\} = \frac{100}{9}e^{-t}u(t) - \frac{100}{9}e^{-10t}u(t)$$

- Find closed-loop step response:

$$G_{CL}(s) = C(s)/R(s) \quad and: u(t) \Leftrightarrow 1/s$$

dc gain?

$$so: C(s) = R(s)G_{CL}(s) = \frac{1}{s}\frac{100}{(s+10)(s+1)} = \frac{10}{s} + \frac{-100/9}{(s+1)} + \frac{10/9}{(s+10)}$$

$$so: c(t) = L^{-1}\left\{\frac{10}{s} + \frac{-100/9}{(s+1)} + \frac{10/9}{(s+10)}\right\} = 10u(t) - \frac{100}{9}e^{-t}u(t) + \frac{10}{9}e^{-10t}u(t)$$

Example Bode Magnitude Plot

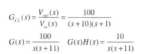

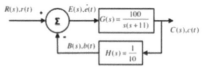

- Plot |G(s)|, |G(s)H(s)|, and closed-loop gain |G$_{CL}$(s)| in dB
  - Recall: Bode magnitude breaks 20dB/decade at 3dB point

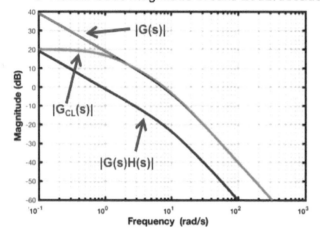

Exercise:

Add the Bode plot analysis lines to the plotted responses

Example: Amplifier Feedback Bode Phase Plot

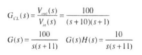

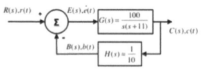

- Plot ∠G(s), ∠G(s)H(s), and closed-loop phase ∠G$_{CL}$(s)
  - Recall: Phase (degrees) is 45° at 3dB point, 45°/decade

Exercise:

Add the Bode plot analysis lines to the plotted responses

3 Z-TRANSFORM

The lecture notes in this chapter review one-sided and two-sided z-transforms, and their application to discrete-time systems.

Discrete-Time Signals

A/D & D/A Conversion, x(t) and x[n]

- ADC: Analog-to-Digital Converter, DAC: Digital-to-Analog Converter
- Consider the ADC/DAC system below (simple A-to-D-to-A system)
- As illustrated, let x[n]=x(nT_0) be the discrete-time signal

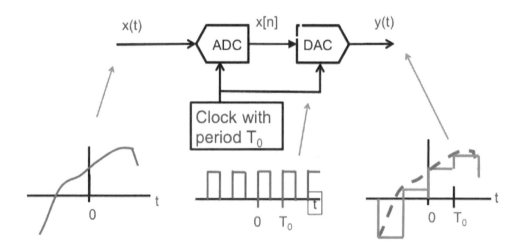

Discrete-Time Signal: x[n]=x(nT₀)

- In sampled system, x [n]= x(n T$_0$)

[ ] Square bracket

( ) parentheses

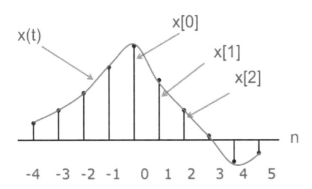

x(t)

x[0]

x[1]

x[2]

n

-4 -3 -2 -1   0   1   2   3   4   5

Discrete-Time Frequency of a Sampled Sinusoid

- Sampled with sampling period T$_0$
- x [n]= x(t)|$_{t=nT_0}$ = x(nT$_0$)
- Example: x(t)=sin(Ω_c t)

$\omega_c = \Omega_c T_0$
discrete-time
frequency

$$x[n] = \sin\left(\frac{2\pi(nT_0)}{T_c}\right) = \sin\left(\frac{2\pi T_0}{T_c} n\right) = \sin(\Omega_c T_0 n) = \sin(\omega_c n)$$

- Ω_c = continuous-time carrier frequency = $2\pi/T_c$ rad/s
- f_c = continuous-time carrier frequency = $1/T_c$ Hz
- Ω_0 = Sampling frequency = $2\pi/T_0$
- f_0 = sampling rate = $1/T_0$ Hz
- $\omega_c = \Omega_c T_0$ discrete-time carrier frequency, rad/sample

$$\omega_c = 2\pi\left(\frac{\Omega_c}{\Omega_0}\right) = \Omega_c T_0 = 2\pi\left(\frac{f_c}{f_0}\right) = 2\pi\left(\frac{T_0}{T_c}\right)$$

Plot of Sampled Sinusoid

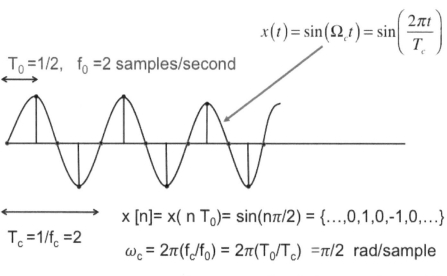

$$x(t) = \sin(\Omega_c t) = \sin\left(\frac{2\pi t}{T_c}\right)$$

$T_0 = 1/2$, $f_0 = 2$ samples/second

$T_c = 1/f_c = 2$

$x[n] = x(n\,T_0) = \sin(n\pi/2) = \{\ldots,0,1,0,-1,0,\ldots\}$

$\omega_c = 2\pi(f_c/f_0) = 2\pi(T_0/T_c) = \pi/2$ rad/sample

How many radians between each sample?

Sampling (Model of A/D+D/A Conversion)

- Consider ADC model below, where h(t) = $\Pi((t-T_0/2) / T_0)$
- $\Pi(t/\tau)$ is rectangular pulse centered at zero of width τ and height 1
- Below, $x_s(t)$ is the sampled signal
- This model has the same input/output as the ADC+DAC system

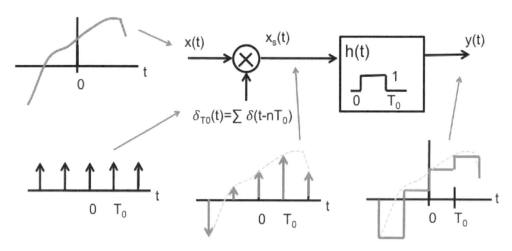

Sampling

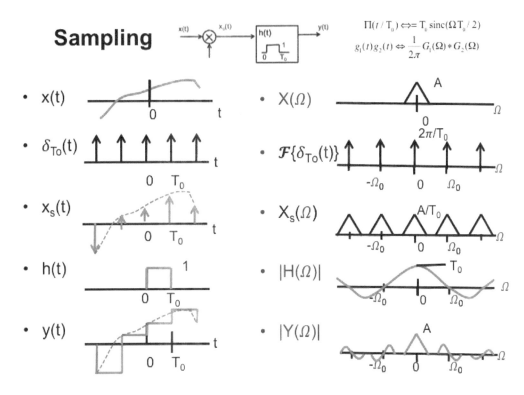

Sampling Redrawn Using Spectra

- This model has the same input/output as the ADC+DAC system
- $h(t) = \Pi\{(t-T_0/2)/(T_0)\}$,

$$G(\Omega) = \int_{-\infty}^{\infty} g(t) e^{-j\Omega t}\, dt$$

$$g(t) = \frac{1}{2\pi} \int_{-\infty}^{\infty} G(\Omega) e^{j\Omega t}\, d\Omega$$

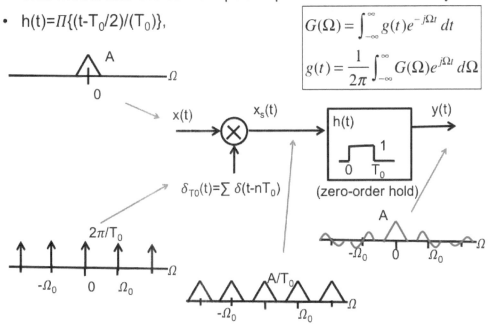

Nyquist Rate and Aliasing

- Consider the spectrum of the sampled signal $X_s(\Omega)$ below
- If $\Omega_0 = 2\pi/T_0$ is too small, the spectra will overlap, and information will be lost
- This overlap is aliasing
- The minimum sampling rate tp prevent aliasing is the Nyquist rate, 2B, where B is the bandwidth of the signal x(t)
- Sampling frequency $f_0 > 2B$ to prevent aliasing, where $f_0 = 1/T_0$

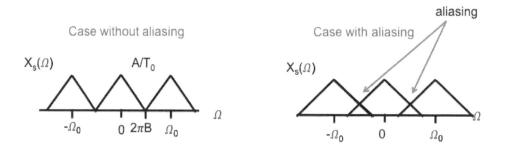

DTFT (Discrete-Time Fourier Transform)

- Recall: $x_s(t) = x(t)\delta_{T0}(t) = x(t) \sum \delta(t-nT_0)$
- So:

$$X_s(\Omega) = \int_{-\infty}^{\infty} x_s(t)e^{-j\Omega t}\, dt$$

$$= \int_{-\infty}^{\infty} \left(\sum_{n=-\infty}^{\infty} x(nT_0)\delta(t-nT_0) \right) e^{-j\Omega t}\, dt$$

$$= \sum_{n=-\infty}^{\infty} x(nT_0) \int_{-\infty}^{\infty} \delta(t-nT_0)e^{-j\Omega t}\, dt$$

$$X_s(\Omega) = \sum_{n=-\infty}^{\infty} x(nT_0)e^{-j\Omega nT_0}$$

x(t) → ⊗ → $x_s(t)$

$\delta_{T0}(t) = \sum \delta(t-nT_0)$

DTFT (Discrete-Time Fourier Transform)

- Now define **the discrete-time Fourier transform** (DTFT)
- Some texts just call this "Fourier transform"

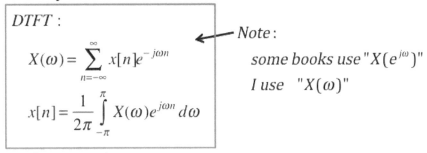

$DTFT:$

$$X(\omega) = \sum_{n=-\infty}^{\infty} x[n]e^{-j\omega n}$$

$$x[n] = \frac{1}{2\pi} \int_{-\pi}^{\pi} X(\omega)e^{j\omega n}\, d\omega$$

← $Note:$

$some\ books\ use\ "X(e^{j\omega})"$

$I\ use\ \ "X(\omega)"$

$by\ comparison:$

$$X_s(\Omega) = \sum_{n=-\infty}^{\infty} x(nT_0)e^{-j\Omega n T_0}$$

$so\ with\ x[n] = x(nT_0)\ and\ with\ \omega = \Omega T_0$

$$X(\omega) = X_s(\Omega)\big|_{\Omega = \omega/T_0}$$

Comments on DTFT

- DTFT is continuous in ω
- DTFT is periodic in ω, with period 2π
- Since $\omega = \Omega T_0 = 2\pi f/f_0$ then: $\omega = 2\pi$ at $f = f_0$
- Result of DTFT is complex
- Closely related to $X_s(\Omega)$

$DTFT:$

$$X(\omega) = \sum_{n=-\infty}^{\infty} x[n]e^{-j\omega n}$$

$$x[n] = \frac{1}{2\pi} \int_{-\pi}^{\pi} X(\omega)e^{j\omega n}\, d\omega$$

$and\ \ X(\omega) = X_s(\Omega)\big|_{\Omega = \omega/T_0}$

$X(\Omega)$ — A — Ω

$X_s(\Omega)$ — A/T_0 — $-\Omega_0$ — Ω_0 — Ω

$X(\omega)$ — A/T_0 — -2π — 2π — ω

DTFT Example

- Find DTFT of 3-point moving sum
  y[n]=x[n] + x[n-1] + x[n-2]
  Note: h[n]=u[n]-u[n-3]

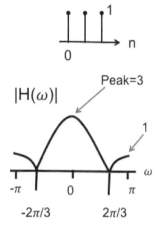

$$H(\omega) = \mathcal{F}\{h[n]\} = \sum_{n=-\infty}^{\infty} h[n]e^{-j\omega n} = \sum_{n=0}^{2} 1 \cdot e^{-j\omega n}$$

$$= \sum_{n=0}^{2}\left(e^{-j\omega}\right)^{n} = \frac{1-\left(e^{-j\omega}\right)^{3}}{1-e^{-j\omega}} = \frac{1-e^{-j3\omega}}{1-e^{-j\omega}}$$

$$= \frac{e^{-j3\omega/2}\left(e^{j3\omega/2}-e^{-j3\omega/2}\right)}{e^{-j\omega/2}\left(e^{j\omega/2}-e^{-j\omega/2}\right)} = \frac{\sin(3\omega/2)e^{-j\omega}}{\sin(\omega/2)}$$

$$\text{using } \sum_{k=N_1}^{N_2} \alpha^k = \frac{\alpha^{N_1}-\alpha^{N_2+1}}{1-\alpha} \quad ; \; N_2 > N_1$$

From the difference equation, if the
input is x[n]=2, what is the output?

Peak=3

|H(ω)|

1

-π   0   π

-2π/3   2π/3

Why does peak=3?
Note: plot is periodic
    Period=?
Is this a lowpass filter?
What is dc response?

DTFT Across Block Diagram

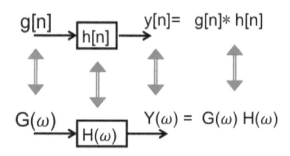

- Take DTFT across block diagram
- Each discrete-time domain element has corresponding frequency domain element
- Two languages have different word for same thing
- Is frequency or time the "true" representation?

Z-Transform

Z- Transform - Simplified

- Just multiply x[n] by z^{-n}

$$X(z) = Z\{x[n]\} = \sum_{-\infty}^{\infty} x[n]z^{-n}$$

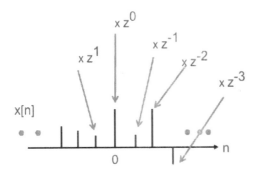

31

One-Sided Z-Transform Definition

- Z-transform is similar to Laplace transform
- The one-sided z-transform is restricted to causal (right-sided) signals x[n]
- Define one-sided z-transform:

$$X(z) = Z\{x[n]\} = \sum_{n=0}^{\infty} x[n]z^{-n}$$

$$x[n] = \frac{1}{2\pi j} \oint X(z)z^{n-1} \, dz$$

1-sided

Closed counter-clockwise
contour integral in ROC

Two-Sided Z-Transform Definition

- Two-sided z-transform is similar to two-sided Laplace transform
- To be valid, two-sided z-transform MUST include **ROC**
- ROC is Region of Convergence of X(z) in z-plane
- The following development is for the more common two-sided z-transform
- Define two-sided z-transform:

$$X(z) = Z\{x[n]\} = \sum_{n=-\infty}^{\infty} x[n]z^{-n}$$

$$x[n] = \frac{1}{2\pi j} \oint X(z)z^{n-1} \, dz$$

2-sided

Closed counter-clockwise
contour integral in ROC

Z- Transform Visualization

Note:

X(z) is a function of z that itself could be complex-valued.

Consider to plot |X(z)| we would need a "3-D plot" where the height above the z-plane would correspond to |X(z)|.

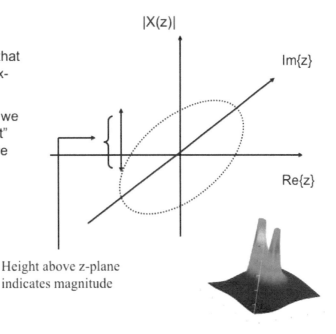

Height above z-plane indicates magnitude

Z- Transform Relation to Laplace Transform

- Z-transform is related to Laplace transform (both 2-sided here)

$$X(z) = Z\{x[n]\} = \sum_{n=-\infty}^{\infty} x[n]z^{-n}$$

$$Define: \quad x_s(t) = \sum_{n=-\infty}^{\infty} x[n]\,\delta(t - nT_0)$$

$$X_s(s) = L\{x(t)\} = \int_{-\infty}^{\infty} \left(\sum_{n=-\infty}^{\infty} x[n]\,\delta(t - nT_0) \right) e^{-st}\,dt$$

Both 2-sided

$$= \sum_{n=-\infty}^{\infty} x[n] \int_{-\infty}^{\infty} \delta(t - nT_0)e^{-st}\,dt = \sum_{n=-\infty}^{\infty} x[n]e^{-snT_0}$$

so :

$$\boxed{X_s(s) = X(z)\big|_{z=e^{sT_0}}}$$

So, $X_s(j\Omega T_s)$ corresponds to points on unit circle in z-plane where $z = e^{j\omega} = e^{j\Omega T_0}$

Discrete-Time Frequency Response H(ω)

- So, z-transform H(z) defines the discrete-time frequency response H(ω)
- H(ω) is most commonly called the DTFT (Discrete-Time Fourier Transform (also often denoted as H($e^{j\omega}$))
- Define:

discrete-time frequency response lies above unit circle

$$Define: \quad H(\omega) = H(z)\big|_{z=e^{j\omega}}$$

- And since:

$$L\{h(t)\} = H_s(s) = H(z)\big|_{z=e^{sT_0}}$$

then :

Both 2-sided

$$\boxed{H(\omega) = H(z)\big|_{z=e^{j\omega}=e^{j\Omega T_0}} = H_s(j\Omega)}$$

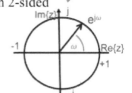

- Note
  - $H(\omega)$ is periodic on ω with period 2π
  - $H_s(j\Omega)$ is periodic in Ω with period $2\pi/T_0$

Example: Z-Transform and DTFT, x[n]={1,1,1,1}

- Example: x[n]=u[n]-u[n-4]

$$X(z) = \sum_{n=-\infty}^{\infty} x[n]z^{-n} = \sum_{n=0}^{3} 1z^{-n} = \frac{1-z^{-4}}{1-z^{-1}}; ROC\ z \neq 0$$

where :

$$\sum_{n=A}^{B} \alpha^n = \frac{\alpha^A - \alpha^{B+1}}{1-\alpha}$$

- Now consider the frequency response on the unit circle
- Denote discrete-time frequency response X(ω) as

Note: many books use X($e^{j\omega}$)

$$X(\omega) = X(z)\big|_{z=e^{j\omega}} = \frac{1-z^{-4}}{1-z^{-1}}\bigg|_{z=e^{j\omega}} = \frac{1-(e^{j\omega})^{-4}}{1-(e^{j\omega})^{-1}} = \frac{1-e^{-j4\omega}}{1-e^{-j\omega}}$$

Also, $|X(\omega)| = |(e^{j\omega 2} - e^{-j2\omega})e^{-j\omega 2}/(e^{j\omega/2} - e^{-j\omega/2})e^{-j\omega/2}|$

$= | e^{-j3\omega/2}2j\sin(2\omega)/2j\sin(\omega/2) |$

$= |\sin(2\omega)/\sin(\omega/2) |$

What is dc response?

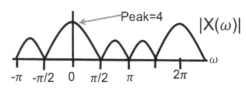

Peak=4

|X(ω)|

$-\pi$ $-\pi/2$ 0 $\pi/2$ π 2π

Example, continued: Poles and Zeroes of H(z)

- From previous result, rearrange

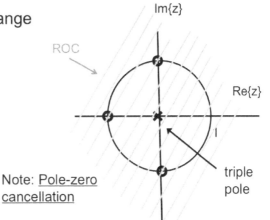

$$H(z) = \frac{1-z^{-4}}{1-z^{-1}} = \frac{z^4-1}{z^3(z-1)}$$

$$= \frac{(z-1)(z-j)(z+1)(z+j)}{z^3(z-1)}$$

$$= \frac{(z-j)(z+1)(z+j)}{z^3};$$

$ROC \quad z \neq 0$

Note: <u>Pole-zero</u> <u>cancellation</u>

- Using long division show also:

$$H(z) = 1 + z^{-1} + z^{-2} + z^{-3}$$

•Note: same poles and zeroes for both forms

Example: Z-Transform of Unit Step

Unit step: x[n] = u[n]

$$X(z) = \sum_{n=-\infty}^{\infty} x[n]z^{-n} = \sum_{n=0}^{\infty} u[n]z^{-n} = \frac{1-z^{-\infty}}{1-z^{-1}}$$

$$X(z) = \frac{1}{1-z^{-1}}; \quad ROC\,|z| > 1$$

•ROC:

$$where: \quad \sum_{n=A}^{B} \alpha^n = \frac{\alpha^A - \alpha^{B+1}}{1-\alpha}$$

•Region of convergence

•Region in z-plane where X(z) is bounded

$$check: X(z)\big|_{z=1} = \sum_{n=-\infty}^{\infty} x[n]z^{-n} = \sum_{n=0}^{\infty} 1 = \infty$$

<u>So</u>

- ROC |z| > 1.

Example: Real Exponential Sequence

- Find the z-transform of $x[n]=a^n u[n]$

Causal ROC

$$X(z)= \sum_{n=-\infty}^{\infty} a^n u[n]z^{-n} = \sum_{n=0}^{\infty} a^n z^{-n} = \sum_{n=0}^{\infty} (az^{-1})^n = \frac{1-(az^{-1})^{+\infty}}{1-az^{-1}}$$

$$= \begin{cases} undefined; & |z| \le a \\ \dfrac{1}{1-az^{-1}}; & |z| > |a| \end{cases}$$

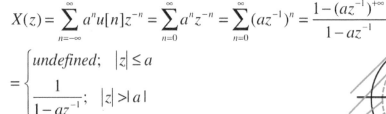

note that if $z = a$;

$$X(z)= \sum_{0}^{\infty} (az^{-1})^n = \sum_{0}^{\infty} (aa^{-1})^n = \sum_{0}^{\infty} 1$$

- <u>Consider</u> the case when a<1
  - $a^n u[n]$ for a=0.9
  - {1,0.9,0.81...}, here a=.9, ROC |z|>.9

In this example (a=0.9) unit circle lies in ROC. (X(ω) exists).

ROC: Region of Convergence for 2-Sided Z-Transform

- ROC: region in z-plane where the z-transform of a sequence exists (is finite).

- ROC is required for 2-sided z-transform

- Convergence (finite value) of z-transform is assured if

$$X(z)= \sum_{n=-\infty}^{\infty} x[n]z^{-n} \le \sum_{n=-\infty}^{\infty} |x[n]z^{-n}| = \sum_{n=-\infty}^{\infty} |x[n]||z^{-n}| = \sum_{n=-\infty}^{\infty} \frac{|x[n]|}{|z|^n} < \infty$$

or using polar coordinates with $z = re^{j\theta}$

$$\sum_{n=-\infty}^{\infty} |x[n]re^{-jn\theta}| = \sum_{n=-\infty}^{\infty} |x[n]||r^{-n}| = \sum_{n=1}^{\infty} |x[-n]||r^n| + \sum_{n=0}^{\infty} \frac{|x[n]|}{|r|^n} < \infty$$

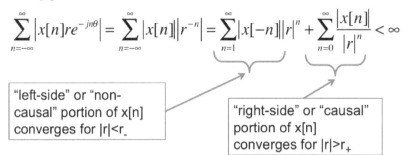

"left-side" or "non-causal" portion of x[n] converges for |r|<r_

"right-side" or "causal" portion of x[n] converges for |r|>r_+

Region of Convergence

$$X(z) = \sum_{n=-\infty}^{\infty} x[n]z^{-n} \le \sum_{n=1}^{\infty}|x[-n]|\,|r|^n + \sum_{n=0}^{\infty}\frac{|x[n]|}{|r|^n} < \infty$$

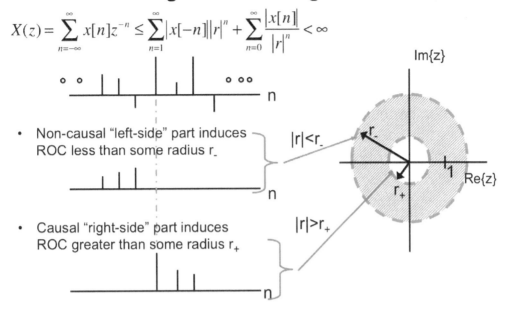

- Non-causal "left-side" part induces ROC less than some radius r_-

  $|r|<r_-$

- Causal "right-side" part induces ROC greater than some radius r_+

  $|r|>r_+$

Total ROC is intersection of both ROC, dashed cyan ring above

2-sided z-transform example

$$x[n] = a^n u[n] - b^n u[-n-1]$$

Left-sided part of sequence

$$X(z) = \sum_{-\infty}^{\infty} x[n]z^{-n} = -\sum_{-\infty}^{-1} b^n z^{-n} + \sum_{n=0}^{\infty} a^n z^{-n}$$

$$= -\sum_{0}^{\infty} b^{-1-n}z^{1+n} + \sum_{n=0}^{\infty} a^n z^{-n}$$

$$= --b^{-1}z\sum_{0}^{\infty}\left(b^{-1}z\right)^n + \sum_{n=0}^{\infty}(az^{-1})^n$$

$$= b^{-1}z\left(\frac{1-(b^{-1}z)^\infty}{1-b^{-1}z}\right) + \frac{1-(az^{-1})^\infty}{1-az^{-1}}$$

$$= -b^{-1}z\left(\frac{1}{1-b^{-1}z}\right) + \frac{1}{1-az^{-1}}$$

$$= \frac{1}{1-bz^{-1}} + \frac{1}{1-az^{-1}}$$

$$ROC\ |z| < |b|\ and\ |z| > |a|$$

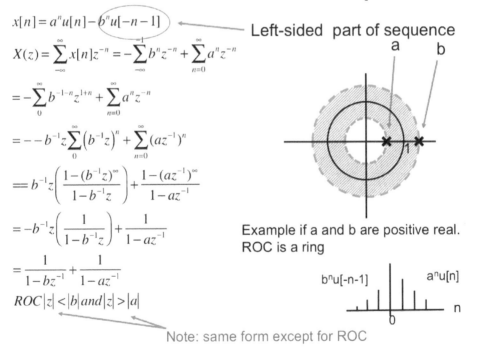

Example if a and b are positive real. ROC is a ring

$b^n u[-n-1]$ $a^n u[n]$

Note: same form except for ROC

Z-Transform MUST Include an ROC

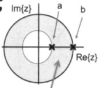

x[n]

- So, for the 2-sided example the left-sided and right-side components of the z-transform had the same form

$$a^n u[n] \Leftrightarrow \frac{1}{1-az^{-1}}; |z| > |a|$$

$$-a^n u[-n-1] \Leftrightarrow \frac{1}{1-az^{-1}}; |z| < |a|$$

Note that poles are found at boundary of ROC

- The only difference in the two forms above are the ROC!!

- So, 2-sided z-transform must include BOTH X(z) and ROC!!

- Not an issue in 1-sided z-transform, since all are right-sided (causal), so ROC is not needed for 1-sided z-trasform

Two-Sided Signal Example

From general form of previous example, choose b=2, a=0.5, and find poles and zeroes:

Left-sided pole at z=b, ROC | z|<|b|=2

$$\frac{z}{z-b} + \frac{z}{z-a} = \frac{z}{z-2} + \frac{z}{z-0.5}$$ Zero @ z=0

$$= \frac{z(z-a) + z(z-b)}{(z-a)(z-b)}$$

$$= \frac{z(2z-a-b)}{(z-a)(z-b)} = \frac{z(2z-2.5)}{(z-0.5)(z-2)}$$

2 zeroes: z=0 & z=(a+b)/2=1.25

Zero @ z=1.25

Causal pole at boundary of causal ROC, pole at z=a ROC |z|>|a|=0.5

Table of Some 2-Sided Z-Transforms
See appendix for $\delta[n]$

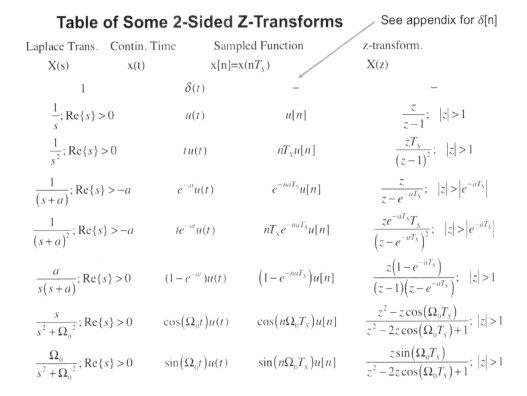

Laplace Trans. $X(s)$	Contin. Time $x(t)$	Sampled Function $x[n]=x(nT_s)$	z-transform. $X(z)$
1	$\delta(t)$	$-$	$-$
$\dfrac{1}{s}$; $\mathrm{Re}\{s\}>0$	$u(t)$	$u[n]$	$\dfrac{z}{z-1}$; $|z|>1$
$\dfrac{1}{s^2}$; $\mathrm{Re}\{s\}>0$	$tu(t)$	$nT_s u[n]$	$\dfrac{zT_s}{(z-1)^2}$; $|z|>1$
$\dfrac{1}{(s+a)}$; $\mathrm{Re}\{s\}>-a$	$e^{-at}u(t)$	$e^{-naT_s}u[n]$	$\dfrac{z}{z-e^{-aT_s}}$; $|z|>\left|e^{-aT_s}\right|$
$\dfrac{1}{(s+a)^2}$; $\mathrm{Re}\{s\}>-a$	$te^{-at}u(t)$	$nT_s e^{-naT_s}u[n]$	$\dfrac{ze^{-aT_s}T_s}{(z-e^{-aT_s})^2}$; $|z|>\left|e^{-aT_s}\right|$
$\dfrac{a}{s(s+a)}$; $\mathrm{Re}\{s\}>0$	$(1-e^{-at})u(t)$	$(1-e^{-naT_s})u[n]$	$\dfrac{z(1-e^{-aT_s})}{(z-1)(z-e^{-aT_s})}$; $|z|>1$
$\dfrac{s}{s^2+\Omega_0^2}$; $\mathrm{Re}\{s\}>0$	$\cos(\Omega_0 t)u(t)$	$\cos(n\Omega_0 T_s)u[n]$	$\dfrac{z^2-z\cos(\Omega_0 T_s)}{z^2-2z\cos(\Omega_0 T_s)+1}$; $|z|>1$
$\dfrac{\Omega_0}{s^2+\Omega_0^2}$; $\mathrm{Re}\{s\}>0$	$\sin(\Omega_0 t)u(t)$	$\sin(n\Omega_0 T_s)u[n]$	$\dfrac{z\sin(\Omega_0 T_s)}{z^2-2z\cos(\Omega_0 T_s)+1}$; $|z|>1$

Properties of Discrete-Time Systems

Discrete-Time System Properties

$$x[n] \rightarrow \boxed{T\{\ \}} \rightarrow y[n] = T\{\ x[n]\ \}$$

1. Linear:

If $y_1[n] = T\{\ x_1[n]\ \}$ and $y_2[n] = T\{\ x_2[n]\ \}$

then, the system is linear if and only if:

$T\{\ ax_1[n] + bx_2[n]\ \} = aT\{\ x_1[n]\ \} + bT\{\ x_2[n]\ \} = ay_1[n] + by_2[n]$

Envision 3 different tests in a laboratory

2. Time/Shift Invariant:

If $y[n] = T\{\ x[n]\ \}$, then $y[n-n_d] = T\{\ x[n-n_d]\ \}\ \forall\ n_d$

Informally, $T\{\ \}$ doesn't change over time

3. BIBO Stability:

BIBO: Bounded Input Bounded Output

Every bounded Input produces bounded output:

If $|x[n]| \le B_x\ \forall\ n$, then $|y[n]| \le B_y\ \forall\ n$

where B_x and B_y are finite bounds

LTI (Linear Time Invariant) System

$$x[n] \longrightarrow \boxed{T\{\ \}} \longrightarrow y[n] = T\{\ x[n]\ \}$$

- Consider imposing linearity and time invariance on $T\{\ \}$
- First. Break x[n] up into sum of many impulses times "weights"

$$x[n] = \sum_{k=-\infty}^{\infty} x[k]\delta[n-k]$$

- Then, apply the transformation $T\{\ \}$:

$$y[n] = T\{x[n]\} = T\left\{ \sum_{k=-\infty}^{\infty} x[k]\delta[n-k] \right\}$$

$$= \sum_{k=-\infty}^{\infty} x[k]T\{\delta[n-k]\} \quad by\ linearity$$

$$= \sum_{k=-\infty}^{\infty} x[k]h[n-k] \quad by\ time\ inv.$$

$$result: \quad y[n] = \sum_{k=-\infty}^{\infty} x[k]h[n-k] = x[n]*h[n] \quad convolution$$

LTI System: Convolution

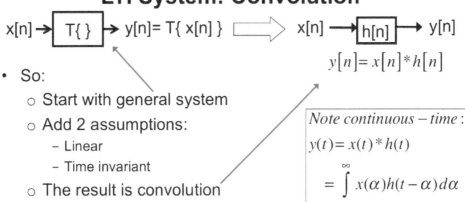

$$y[n]= x[n]*h[n]$$

- So:
  - Start with general system
  - Add 2 assumptions:
    - Linear
    - Time invariant
  - The result is convolution

$$Note\ continuous-time:$$
$$y(t)= x(t)*h(t)$$
$$= \int_{-\infty}^{\infty} x(\alpha)h(t-\alpha)d\alpha$$

- LTI is "common sense" behavior
  - Linear: knock on door twice as hard, the sound is twice as loud
  - Time invar.: knock on door tomorrow will make sound same as today
- Also, note:

$$h[n]*x[n]= \sum_{k=-\infty}^{\infty} h[k]x[n-k]= \sum_{k=-\infty}^{\infty} x[k]h[n-k]= x[n]*h[n]$$

Convolution: Sum of weighted- delayed h[n]'s

- Output y[n] = sum of weighted-delayed impulse responses

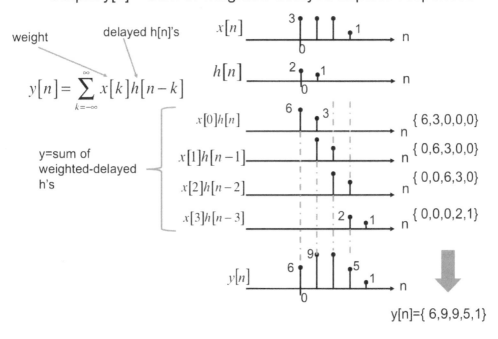

$$y[n]=\{ 6,9,9,5,1\}$$

Properties of z-transform

Properties of Two-Sided Z-Transform

1. Linearity (true for 1-sided, too):

$$ax_1[n]+bx_2[n] \Leftrightarrow aX_1(z)+bX_2(z)$$

Text uses "T" in blocks for delay

2. Shift property of 2-sided z-transform:

$$Z\{x[n-n_o]\} = \sum_{n=-\infty}^{\infty} x[n-n_o]z^{-n}$$

$$= \sum_{\alpha=-\infty}^{\infty} x[\alpha]z^{-(\alpha+n_o)} = z^{-n_o}\sum_{-\infty}^{\infty}x[\alpha]z^{-\alpha}$$

$$= z^{-n_o}X(z)$$

$$Z\{x[n-n_o]\} \Leftrightarrow z^{-n_o}X(z)$$

Beware: for 1-sided z-transform, above is only valid for $n_0 \geq 0$

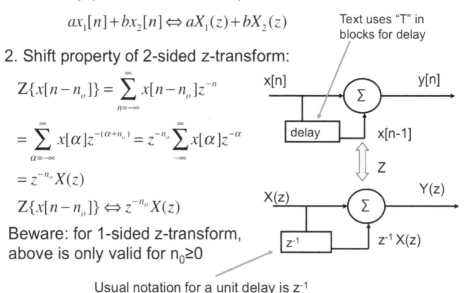

Usual notation for a unit delay is z^{-1}

Shift Property of One-Sided Z-Transform

- Recall, 1-sided z-transform presumes all signals are causal
- For 1-sided z-transform, shift-left is then a problem
- So, shift-property for 1-sided z-transform:

$$Z\{x[n+n_1]\} = \sum_{n=0}^{\infty} x[n+n_1]z^{-n} = z^{n_1}\sum_{\alpha=n_1}^{\infty} x[\alpha]z^{-\alpha}$$

> Removes values to left of n=0

$$= z^{n_1}\sum_{n=0}^{\infty} x[n]z^{-n} - z^{n_1}\sum_{n=0}^{n_1-1} x[n]z^{-n} = z^{n_1}X(z) - z^{n_1}\sum_{n=0}^{n_1-1} x[n]z^{-n}$$

summarizing:

$$Z\{x[n-n_o]\} = z^{-n_o}X(z); \quad for \ n_o \geq 0 \quad \text{right-shift property}$$

$$Z\{x[n+n_1]\} = z^{n_1}X(z) - z^{n_1}\sum_{n=0}^{n_1-1} x[n]z^{-n}; \quad for \ n_1 > 0 \quad \text{left-shift property}$$

Beware: for 1-sided z-transform, the left-shifted data "disappears" to the left of n=0

Convolution Property of 2-sided z-transform

3. Convolution:

$$Z\{x[n]*h[n]\} = \sum_{n=-\infty}^{\infty}\left(\sum_{k=-\infty}^{\infty} x[k]h[n-k]\right)z^{-n}$$

y[n]= x[n]*h[n]

x[n]

$$= \sum_{k=-\infty}^{\infty} x[k]\left(\sum_{n=-\infty}^{\infty} h[n-k]z^{-n}\right)$$

h[n]

$$= \sum_{k=-\infty}^{\infty} x[k]z^{-k}H(z) = H(z)\sum_{k=-\infty}^{\infty} x[k]z^{-k}$$

Z

$$= H(z)X(z)$$

X(z)

Y(z)= X(z)H(z)

H(z)

$$so \quad x[n]*h[n] \Leftrightarrow X(z)H(z)$$

- ROC= intersection of ROCs typically.

43

Convolution Property of One-Sided Z-Transform

- Recall, 1-sided z-transform presumes all signals are causal
- So 1-sided z-transform, convolution sum begins at n=0
- So, convolution-property for 1-sided z-transform:

$$Z\{x[n]*h[n]\} = \sum_{n=0}^{\infty}\left(\sum_{k=0}^{\infty}x[k]h[n-k]\right)z^{-n} = \sum_{k=0}^{\infty}x[k]\left(\sum_{n=0}^{\infty}h[n-k]z^{-n}\right)$$

$$= \sum_{k=0}^{\infty}x[k]z^{-k}H(z) = H(z)\sum_{k=0}^{\infty}x[k]z^{-k} = X(z)H(z)$$

Sums begin at n=0 for causal 1-sided z-transform

summarizing:

$$Z\{x[n]*h[n]\} = X(z)H(z); \quad \text{for any h[n] and x[n]}$$

and

$$Z_{1-sided}\{(x[n]u[n])*(h[n]u[n])\} = Z_{2-sided}\{(x[n]u[n])*(h[n]u[n])\}$$

- Beware: convolution sums are different!!
- Only equal if both functions are causal

Properties of z-transform

4. Exponential modulation property:

$$z_o^n x[n] \Leftrightarrow X(z/z_o)$$

$$Z\{z_o^n x[n]\} = \sum_{n=-\infty}^{\infty} z_o^n x[n]z^{-n}$$

$$= \sum_{n=-\infty}^{\infty} x[n](z^{-1}z_o)^n = \sum_{n=-\infty}^{\infty} x[n]\left(\frac{z}{z_o}\right)^{-n}$$

$$= X(z)\big|_{z=z/z_o} = X(z/z_o)$$

In particular,

$$\left(e^{j\omega_o}\right)^n x[n] \Leftrightarrow X(z/e^{j\omega_o})$$

Rotation in z-plane

where

This is z-domain version of frequency-shift property

$$X\left(\frac{z}{e^{j\omega_o}}\right) = X\left(re^{j\omega}e^{-j\omega_o}\right) = X\left(re^{j(\omega-\omega_o)}\right)$$

Properties of z-transform

5. BIBO Stability (true for 1-sided, too):
   - BIBO **stable if the ROC includes unit circle**,
   - For 1-sided z-transform, all poles must be inside unit circle

6. Also, note in time domain, BIBO stable if

$$\sum_{k=-\infty}^{\infty} x[k]h[n-k] \le \sum_{k=-\infty}^{\infty} |B_x| |h[n-k]| \le \infty \Rightarrow \sum_{k=-\infty}^{\infty} |h[k]| < \infty$$

7. Initial and final value (true for 1-sided, too):

$$\lim_{z \to \infty} X(z) = x[0] \quad \text{for causal x[n] only}$$

$$\lim_{z \to 1} \left(1 - z^{-1}\right) X(z) = x[\infty] \quad \text{for causal x[n] only}$$

Block Diagrams and Difference Equations

- Difference equation for a block diagram

$$y[n] = -a_1 y[n-1] - a_2 y[n-2] \ldots + b_0 x[n] + b_1 x[n-1] + \ldots$$

- Small triangles indicate weights/gains (multipliers)
- z^{-1} blocks represent unit delay (shift register)

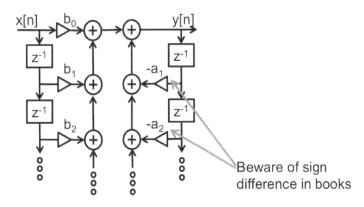

Beware of sign difference in books

Block Diagram example

- Find the difference equation and H(z) for this block diagram

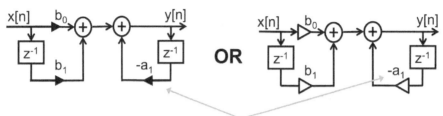

OR

Beware of sign here

- Difference equation: y[n]= -a$_1$y[n-1]+b$_0$x[n]+b$_1$x[n-1]

- Extensions of the above form yield general implementations of difference equations and H(z)

- For above example:

$$H(z) = \frac{Y(z)}{X(z)} = \frac{b_0 + b_1 z^{-1}}{1 + a_1 z^{-1}}$$

Block Diagrams of Discrete-Time Systems

$$y[n] = -a_1 y[n-1] - a_2 y[n-2]... + b_0 x[n] + b_1 x[n-1] + ...$$

z-transform

$$Y(z) = -a_1 z^{-1} Y(z) - a_2 z^{-2} Y(z)... + b_0 X(z) + b_1 z^{-1} X(z) + ...$$

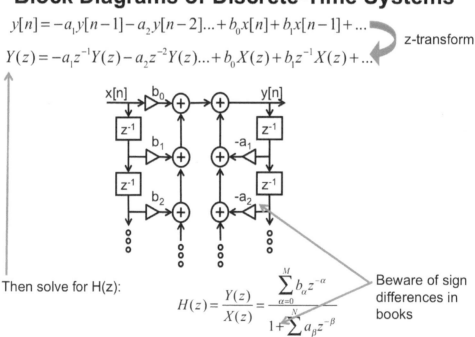

Then solve for H(z):

$$H(z) = \frac{Y(z)}{X(z)} = \frac{\sum_{\alpha=0}^{M} b_\alpha z^{-\alpha}}{1 + \sum_{\beta=1}^{N} a_\beta z^{-\beta}}$$

Beware of sign differences in books

Significance of X(z) at z=1

- X(z) at z=1 corresponds to dc

- Recall relation to DTFT

$$X(\omega) = X(z)\big|_{z=e^{j\omega}}$$

- It is simple to compute X(1):

$$X(1) = X(z)\big|_{z=1}$$

$$X(1) = \sum_{n=-\infty}^{\infty} x[n]z^{-n}$$

$$= \sum_{n=-\infty}^{\infty} x[n]$$

What DTFT frequency ω corresponds to z=1?

z=-1?

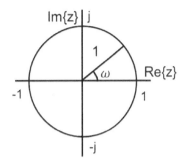

example:

$$X(z) = \frac{z}{z-0.1}; \quad \text{ROC } |z|>0.1$$

$$X(1) = \frac{1}{1-0.1} = \frac{1}{0.9}$$

Example Block Diagram and H(z)

- Example Z-transform and difference equation from block diagram

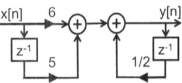

$$y[n] = y[n-1]/2 + 6x[n] + 5x[n-1]$$

$$Y(z) = z^{-1}Y(z)/2 + 6X(z) + 5z^{-1}X(z)$$

$$H(z) = \frac{Y(z)}{X(z)} = \frac{6+5z^{-1}}{1-z^{-1}/2} = \frac{6z+5}{z-1/2}; \quad |z| > 0.5 \quad \longleftarrow \quad \text{Add causal ROC for 2-sided}$$

- Impulse response: from diagram, assume registers "at rest"
- Then h[n]= {6, 8, 4, 2, ...}
- Stable because $\sum|h[n]| < \infty$ and poles inside unit circle

Example: z/(z-a)

- Block diagram can be derived from H(z):

$$\frac{Y(z)}{X(z)} = H(z) = Z\{a^n u[n]\} = \frac{1}{1-az^{-1}} = \frac{z}{z-a}; \quad ROC|z|>|a|$$

$$so: \quad Y(z)(1-az^{-1}) = X(z) \Rightarrow Y(z) = X(z) + a\,z^{-1}Y(z)$$

•What is difference eq.?

$$y[n] = x[n] + ay[n-1]$$

•Is it stable?

•From block diagram, h[n]={1, a, a², ...}

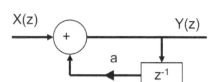

Example: Accumulator

- Accumulator (running sum):

$$y[n] = \sum_{\alpha=-\infty}^{n} x[\alpha]$$

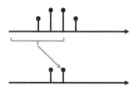

- What is y[n] when input x[n] is an impulse?
- What is y[n] when x[n] = u[n]?
- Is this system linear?
- Is it time invariant?
- What is the impulse response h[n]?
- Is this system BIBO stable?
- Is the block diagram correct?

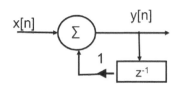

Example: Backward Difference

- Backward difference: y[n]=x[n]-x[n-1] so Y(z)=(1-z⁻¹)X(z)
- So: H(z)=Y(z)/X(z)=1-z⁻¹, and the magnitude of frequency response is

$$H(\omega) = H(z)\Big|_{z=e^{j\omega}} = \left|1 - e^{-j\omega}\right| = \left|\left(e^{j\omega/2} - e^{-j\omega/2}\right)e^{-j\omega/2}\right| = 2\left|\sin(\omega/2)\right|$$

- Convolve the following two functions to find the system output, y[n]
-                x[n]= {1,2,3,4} (ramp)
-                h [n]= {1,-1}   (backward difference)
- Is this an LTI system?
- Does it approximate the derivative?
- What is the block diagram?

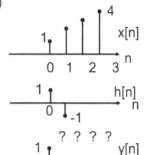

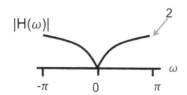

Why does peak |H()|=2?
Note: plot is periodic
Period=?
Is this a lowpass filter?
What is dc response?

Example: H(z) from Difference Equation

x[n]   y[n]

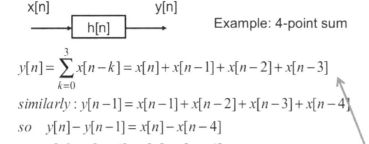

Example: 4-point sum

$$y[n] = \sum_{k=0}^{3} x[n-k] = x[n] + x[n-1] + x[n-2] + x[n-3]$$

$$similarly: y[n-1] = x[n-1] + x[n-2] + x[n-3] + x[n-4]$$

$$so \quad y[n] - y[n-1] = x[n] - x[n-4]$$

$$or \quad y[n] = y[n-1] + x[n] - x[n-4]$$

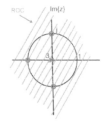

Difference equation is NOT unique for same system h[n]

Take z-transform of both sides:

$$Y(z) - z^{-1}Y(z) = X(z) - z^{-4}X(z) \quad or \quad Y(z)\left(1 - z^{-1}\right) = \left(1 - z^{-4}\right)X(z)$$

$$Y(z) = \frac{1 - z^{-4}}{1 - z^{-1}}X(z) = H(z)X(z) \quad so \quad H(z) = \frac{1 - z^{-4}}{1 - z^{-1}}; \ ROC|z| > 0$$

- For causal solution ROC must be outside outermost pole

- Same as before has pole-zero cancellation.

Inverse z-transform

Inverse Z-Transform Methods

1) Contour integral method: contour in ROC (outside outermost pole for 1-sided z-transform)

$$x[n] = \frac{1}{2\pi j} \oint_C X(z) z^{n-1} \, dz$$

$$recall: \quad \oint_C \frac{G(s)}{(s+a)^n} \, ds = \begin{cases} j2\pi G(-a) & for \ n=1 \\ 0 & otherwise \end{cases}$$

Example contour for 2-sided ROC

$so:$

$$\frac{1}{j2\pi} \oint_C X(z) z^{n-1} \, dz = \frac{1}{j2\pi} \oint_C \frac{X(z) z^n}{z} \, dz = \frac{1}{j2\pi} \oint_C \frac{z^n \sum_{k=-\infty}^{\infty} x[k] z^{-k}}{z} \, dz$$

$$= \frac{1}{j2\pi} \left(\oint_C \sum_{\alpha=-\infty}^{0} \frac{x[n+\alpha-1]}{(z-0)^\alpha} \, dz + \oint_C \frac{x[n]}{(z-0)^1} \, dz + \oint_C \sum_{\beta=2}^{\infty} \frac{x[n+\beta-1]}{(z-0)^\beta} \, dz \right) = x[n]$$

$$So \Rightarrow x[n] = \frac{1}{j2\pi} \oint_C X(z) z^{n-1} \, dz$$

50

Inverse Z-Transform Methods

2) Lookup in z-transform table, including the ROC (will later discuss partial fraction expansion for lookup method)
3) Power series expansion method (long division method)

- Long division example inverse z-transform of right-sided X(z)

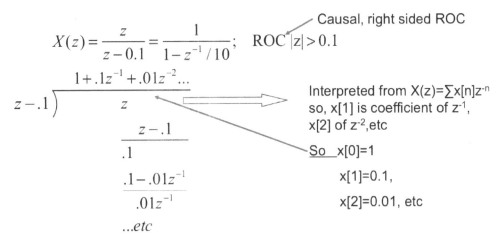

Causal, right sided ROC

$$X(z) = \frac{z}{z - 0.1} = \frac{1}{1 - z^{-1}/10}; \quad ROC\ |z| > 0.1$$

Interpreted from $X(z) = \sum x[n]z^{-n}$
so, x[1] is coefficient of z^{-1},
x[2] of z^{-2}, etc

So x[0]=1

x[1]=0.1,

x[2]=0.01, etc

Inverse Z-Transform Long Division, Left-Sided

$$X(z) = \frac{z}{z - 0.1} = \frac{1}{1 - z^{-1}/10}; \quad ROC\ |z| < 0.1$$

- Example long-division inverse z-transform of left-sided X(z)

- Since ROC |z|<1, then use left sided solution below

$$-.1 + z \overline{) \begin{array}{l} -10z - 100z^2 ... \\ z \\ \underline{z - 10z^2} \\ 10z^2 \\ \underline{10z^2 - 100z^3} \\ 100z^3 \\ ...etc \end{array}}$$

Change order

So:

x[-1]=-10

x[-2]=-100

etc

Inverse Z-Transform Long Division, FIR

- Example inverse z-transform of right-sided X(z)

$$X(z) = \frac{1-z^{-4}}{1-z^{-1}} = \frac{z^4-1}{z^4-z^3}; \quad |z|>0 \qquad \underline{\text{Causal ROC}}$$

Causal long division form

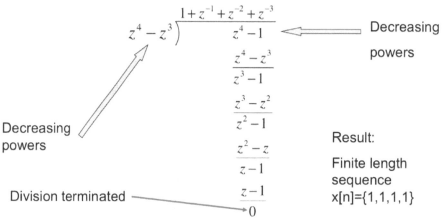

Decreasing powers

Decreasing powers

Division terminated

Result:

Finite length sequence
x[n]={1,1,1,1}

Inverse Z-Transform by Partial Fraction Expansion

- **Consider partial fraction expansion for first-order poles**

   1) Suppose X(z)=N(z)/D(z),

   - Where degree of N(z) < degree of D(z)

   - If not: First do long division until the degree of N(z) is less than the degree of D(z)

   2) Factor denominator, and set N(z) = zQ(z)

$$X(z) = \frac{N(z)}{D(z)} = \frac{z\,Q(z)}{D(z)}$$

$$= \frac{z\,Q(z)}{(z-p_1)(z-p_2)\circ\circ\circ(z-p_i)} = z\sum_{\alpha=1}^{i}\frac{c_\alpha}{z-p_\alpha}$$

Partial Fraction Expansion

From previous slide

$$\frac{Q(z)}{(z-p_1)(z-p_2)\circ\circ\circ(z-p_i)} = \sum_{\alpha=1}^{i} \frac{c_\alpha}{z-p_\alpha}$$

3. Multiply both sides by (z-p$_k$)

$$(z-p_k)\frac{Q(z)}{D(z)} = \frac{(z-p_k)Q(z)}{(z-p_1)(z-p_2)\circ\circ\circ(z-p_i)} = (z-p_k)\left\{\sum_{\alpha=1}^{i}\frac{c_\alpha}{z-p_\alpha}\right\}$$

4. At z=p$_k$ solve for c$_k$

$$(z-p_k)\frac{Q(z)}{D(z)}\bigg|_{z=p_k} = c_k$$

Partial Fraction Expansion

From previous slide

6. Having solved for c$_k$ the partial fraction expansion is:

$$X(z) = \frac{N(z)}{D(z)} = z\frac{Q(z)}{D(z)} = z\sum_{\alpha=1}^{i}\frac{c_\alpha}{z-p_\alpha} = \sum_{\alpha=1}^{i}\frac{zc_\alpha}{z-p_\alpha}$$

7. Taking inverse z-transform

$$a^n u[n] \Longleftrightarrow \frac{z}{z-a}; \quad |z|>|a|$$

$$So: \quad x[n] = Z^{-1}\{X(z)\} = Z^{-1}\left\{\sum_{\alpha=1}^{i}\frac{zc_\alpha}{z-p_\alpha}\right\} = \sum_{\alpha=1}^{i}c_\alpha(p_\alpha)^n u[n]$$

Note: method in text does partial fraction expansion in powers of z^{-1}

Partial Fraction Expansion Example

- <u>Partial Fraction Expansion Example:</u>

$$X(z) = \frac{1}{(z-1/2)(z-1/4)} = z\frac{Q(z)}{D(z)} = z\frac{1}{(z-1/2)(z-1/4)z}; \quad ROC\,|z|>1/2$$

$$= z\left(\frac{\left.\frac{(z-1/2)Q(z)}{D(z)}\right|_{z=1/2}}{z-1/2} + \frac{\left.\frac{(z-1/4)Q(z)}{D(z)}\right|_{z=1/4}}{z-1/4} + \frac{\left.\frac{(z)Q(z)}{D(z)}\right|_{z=0}}{z} \right)$$

$$= z\left(\frac{\left.\frac{1}{(z-1/4)z}\right|_{z=1/2}}{z-1/2} + \frac{\left.\frac{1}{(z-1/2)z}\right|_{z=1/4}}{z-1/4} + \frac{\left.\frac{1}{(z-1/4)(z-1/2)}\right|_{z=0}}{z} \right) = \frac{8z}{z-1/2} - \frac{16z}{z-1/4} + 8$$

- ROC indicates causal right-sided x[n]

- No need for long division since numerator degree is less than denominator

Example, continued

- Rearrange partial fraction expansion results into forms where x[n] may be found using z-transform tables

- Use z-transform properties tables, as needed

- For preceding example:

$$= \frac{8z}{z-1/2} - \frac{16z}{z-1/4} + 8$$

$so:$

$$x[n] = 8\left(\frac{1}{2}\right)^n u[n] - 16\left(\frac{1}{4}\right)^n u[n] + 8\delta[n]$$

Partial Fraction Expansion Example (powers of z⁻¹)

- Same problem(using powers of z^{-1} instead of powers of z)

$$X(z) = \frac{1}{(z-1/2)(z-1/4)} = \frac{z^{-2}}{\left(1-\frac{1}{2}z^{-1}\right)\left(1-\frac{1}{4}z^{-1}\right)} = \frac{z^{-2}}{1-\frac{3}{4}z^{-1}+\frac{1}{8}z^{-2}}$$

Long division first, since denominator order = numerator order

$$+\frac{1}{8}z^{-2}-\frac{3}{4}z^{-1}+1 \,\overline{)z^{-2}}^{\,+8}$$

$$\frac{z^{-2}-6z^{-1}+8}{6z^{-1}-8}$$

So, taking remainder, and doing expansion in powers of z^{-1} :

$$X(z) = 8 + \frac{-8+6z^{-1}}{\left(1-\frac{1}{2}z^{-1}\right)\left(1-\frac{1}{4}z^{-1}\right)} = 8 + \frac{-8+6z^{-1}}{1-(1/4)z^{-1}}\bigg|_{z^{-1}=2} + \frac{-8+6z^{-1}}{1-(1/2)z^{-1}}\bigg|_{z^{-1}=4}$$

$$= 8 + \frac{8}{1-\frac{1}{2}z^{-1}} - \frac{16}{1-\frac{1}{4}z^{-1}}$$

Same result as before!

Bilinear Transform Filter Design Method

- The bilinear transform filter design method avoids aliasing by "squeezing" $-\infty < \Omega < \infty$ into $-\pi < \omega < \pi$
- Given continuous-time filter $H_c(s)$, the bilinear transform yields:

$$H(z) = H_c(s)\bigg|_{s=\frac{2}{T_0}\frac{z-1}{z+1}=\frac{2}{T_0}\frac{1-z^{-1}}{1+z^{-1}}} \quad where \; \frac{1}{T_0} \; is \; the \; sample \; rate$$

- Thus, simply replace "s" in $H_c(s)$ by $(2/T_0)(z-1)/(z+1)$

- Important aspects of bilinear transform
  - No aliasing
  - Stable continuous-time filter guarantees stable H(z)
  - Results in a frequency-warped filter response
- Butterworth filter example

$$H(z) = \frac{\Omega_c^2}{s^2+s\Omega_c\sqrt{2}+\Omega_c^2}\bigg|_{s=\frac{2}{T_0}\frac{z-1}{z+1}} = \frac{\Omega_c^2}{\left(\frac{2}{T_0}\frac{z-1}{z+1}\right)^2+\Omega_c\sqrt{2}\left(\frac{2}{T_0}\frac{z-1}{z+1}\right)+\Omega_c^2}$$

4 STARRED TRANSFORM

The lecture notes in this chapter present the starred transform and discuss the application of the starred transform to the design and analysis of digital systems.

Starred Transform

A/D & D/A Conversion, x(t) and x[n]

- Again, consider the ADC/DAC system below (simple A-to-D-to-A system)
- As illustrated, let $x[n]=x(nT_0)$ be the discrete-time signal

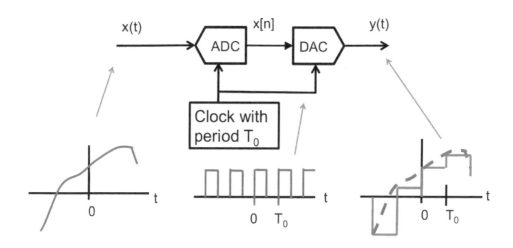

Ideal Sampler and ZOH (Model of A/D+D/A Conversion)

- Again, consider ADC model below, where h(t) = h(t) $=\Pi((t-T_0/2) / T_0)$
- $\Pi(t/\tau)$ is is rectangular pulse centered at zero of width τ and height 1
- Below, $x^*(t)$ is the sampled signal (or ideal-sampled signal)
- This model has the same input/output as the ADC+DAC system
- The h(t) below is a ZOH (zero-order hold)

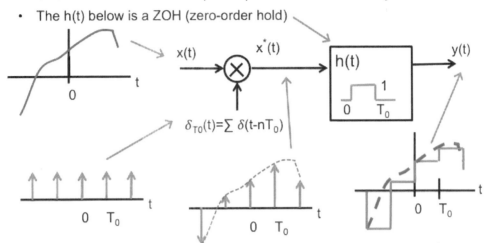

Sampling Redrawn Using Spectra

- This model has the same input/output as the ADC+DAC system
- $h(t)=\Pi\{(t-T_0/2)/(T_0)\}$,

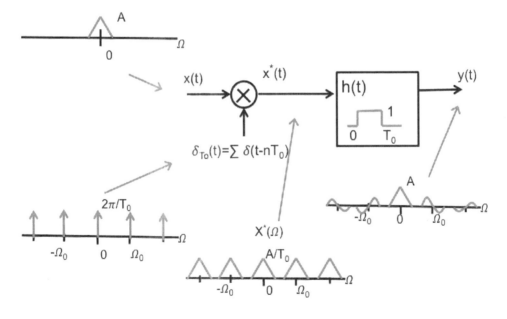

Nyquist Rate and Aliasing

- Consider the spectrum of the sampled signal, $X^*(\Omega)$ below
- If $\Omega_0 = 2\pi/T_0$ is too small, the spectra will overlap, and information will be lost
- This overlap is aliasing
- The minimum sampling rate to prevent aliasing is the Nyquist rate, 2B, where B is the bandwidth of the signal x(t) in Hz
- So $f_0 > 2B$ to prevent aliasing, where $f_0 = 1/T_0$

Case without aliasing

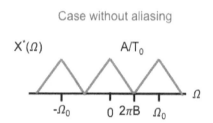

Case with aliasing

aliasing

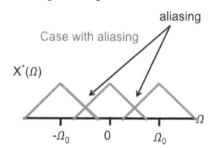

Perfect Reconstruction Filter

- The filter that modeled the ADC-DAC pair had $h(t) = \Pi((t-T_0/2) / T_0)$, and is referred to as a ZOH (zero-order hold) filter
- This filter had the effect of distorting the time domain signal by changing the smooth x(t) into a "staircase/stepwise" approximation y(t)
- In the frequency domain, $H(\Omega)$ multiplied the signal spectrum by a sinc() function, and did not remove the harmonics from the spectrum
- Below, if h(t)is an ideal lowpass filter with $H(\Omega) = T_0 \, \Pi(\Omega/\Omega_0)$, then the output of the system will exactly equal the original input; y(t) = x(t) and $Y(\Omega) = X(\Omega)$

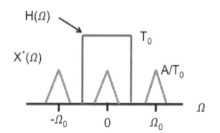

Perfectly reconstructed signal

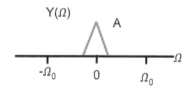

Sampling Theorem

- Nyquist-Shannon Sampling theorem
  - If a signal is strictly bandlimited to B Hz, then it can be perfectly reconstructed from its samples, if the signal is sampled at sampling rate greater than 2B Hz.
- This means that all time points between the sample points can be exactly recovered
- This theorem is essentially the same result as the perfect reconstruction systems given in the previous slides, since y(t) = x(t) !!

Starred Transform X*(s) and ZOH Output

- Laplace transform of sampled signal x*(t) is the **starred transform** X*(s)

$$x^*(t) = x(t)\delta_{T_0}(t) = \sum_{n=-\infty}^{\infty} x(nT_0)\delta(t - nT_0) \quad \Leftrightarrow \quad \boxed{X^*(s) = \sum_{n=-\infty}^{\infty} x(nT_0)e^{-nsT_0}}$$

- The Laplace transform of ZOH system output $x_0(t)$ is:

$$x_0(t) = x^*(t) * \Pi\left(\frac{t - T_0/2}{T_0}\right) \quad \Leftrightarrow \quad X_0(s) = X^*(s)\left(\frac{1 - e^{-sT_0}}{s}\right)$$

Starred transform generally deals with causal signals and 1-sided Laplace transforms

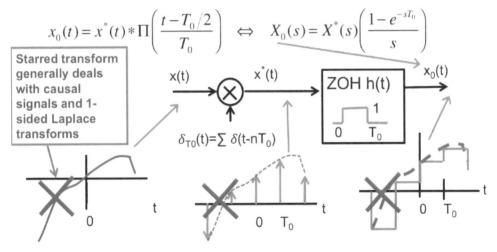

$x(t)$ ⊗ $x^*(t)$ ZOH h(t) $x_0(t)$

$\delta_{T0}(t) = \sum \delta(t - nT_0)$

Note on Starred Transform and Causality

- The starred transform is **restricted to causal** signals x(t)
- **Henceforth**, when we use starred transform, the time function is implicitly causal and may be arbitrarily multiplied by u(t)
- **Henceforth**, when we relate a 2-sided z-transform to a starred transform, the corresponding discrete function must be causal
- With these causal interpretations of starred transforms, derived results are the same whenever 2-sided or 1-sided Laplace transforms or z-transforms are used in derivations. So, for causal x(t):

$$x^*(t) = x(t)\delta_{T_0}(t) = x(t)u(t)\delta_{T_0}(t) = x(t)u(t)\sum_{n=-\infty}^{\infty}\delta(t - nT_0)$$

$$x(t)\sum_{n=0}^{\infty}\delta(t - nT_0) = \sum_{n=0}^{\infty}x(nT_0)u(nT_0)\delta(t - nT_0)$$

> **Starred transform is restricted to causal signals**

$$x^*(t) \quad\Leftrightarrow\quad X^*(s) = \sum_{n=0}^{\infty}x(nT_0)e^{-nsT_0} = Z\{x[n]u[n]\}\big|_{z=e^{sT_0}}$$

> **Very few authors make use of a bilateral starred transform, even though bilateral z-transforms are most common in the signal processing field**

Spectra of Starred Transform

> Starred transform generally deals with causal signals and 1-sided Laplace transforms

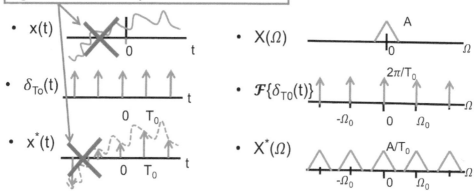

- x(t)
- $\delta_{T_0}(t)$
- $x^*(t)$
- $X(\Omega)$
- $\mathcal{F}\{\delta_{T_0}(t)\}$
- $X^*(\Omega)$

> **DO NOT confuse starred transform notation with complex conjugate**

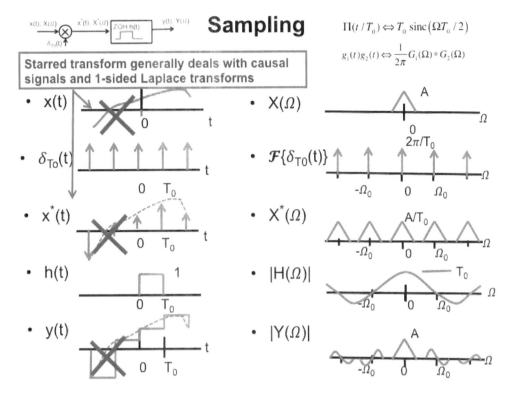

Sampling

$$\Pi(t/T_0) \Leftrightarrow T_0 \operatorname{sinc}(\Omega T_0/2)$$

$$g_1(t)g_2(t) \Leftrightarrow \frac{1}{2\pi}G_1(\Omega) * G_2(\Omega)$$

Starred transform generally deals with causal signals and 1-sided Laplace transforms

- $x(t)$
- $\delta_{T0}(t)$
- $x^*(t)$
- $h(t)$
- $y(t)$

- $X(\Omega)$
- $\mathcal{F}\{\delta_{T0}(t)\}$
- $X^*(\Omega)$
- $|H(\Omega)|$
- $|Y(\Omega)|$

Perfect Reconstruction System

- Consider model below, where $H(\Omega) = T_0\, \Pi(\Omega/\Omega_0)$
- Now the output signal $y(t)$ exactly equals the input signal $x(t)$

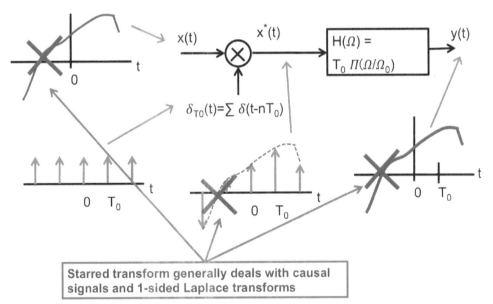

$x(t)$ $x^*(t)$ $H(\Omega) = T_0\, \Pi(\Omega/\Omega_0)$ $y(t)$

$\delta_{T0}(t) = \sum \delta(t-nT_0)$

Starred transform generally deals with causal signals and 1-sided Laplace transforms

Perfect Reconstruction (Convolution View)

- The impulse response of the perfect reconstruction filter is h(t)=sinc(πt/T$_0$)
- When convolved with x*(t), the sinc() functions interpolate between samples as illustrated below

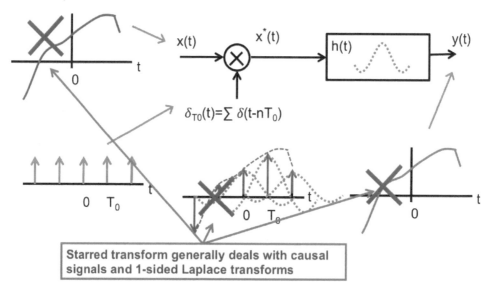

Starred transform generally deals with causal
signals and 1-sided Laplace transforms

Laplace Transform of ZOH

ZOH h(t)

- Laplace transform of sampled signal ZOH is most readily derived by first defining h(t) in terms of unit step functions u(t):

$$h(t) = \Pi\left(\frac{t - T_0/2}{T_0}\right) = u(t) - u(t - T_0)$$

- Then the Laplace transform of ZOH is:

Laplace transform
of ZOH

$$H(s) = L\{u(t) - u(t - T_0)\} = \frac{1 - e^{-sT_0}}{s}$$

$$where:$$

$$u(t) \Leftrightarrow \frac{1}{s}$$

Relation of Starred Transform to Z-Transform

- From previous slide, the starred transform of x(t) is $X^*(s)$

$$X^*(s) = \sum_{n=0}^{\infty} x(nT_0)e^{-nsT_0} = \sum_{n=-\infty}^{\infty} x(nT_0)u(nT_0)e^{-nsT_0}$$

> **Recall:**
> **Starred transform is restricted to causal signals**

- The z-transform of the data samples is

$$X(z) = \sum_{n=0}^{\infty} x(nT_0)z^{-n} = \sum_{n=-\infty}^{\infty} x[n]u[n]z^{-n}$$

- So the relation between z-transform and starred transform $X^*(s)$ is

$$X^*(s) = \sum_{n=0}^{\infty} x(nT_0)e^{-nsT_0} = X(z)\Big|_{z=e^{sT_0}}$$

- So the starred transform $X^*(s)$ is:
  - Is periodic with period $j\omega = sT_0 = j\Omega T_0 = j2\pi$ or every $\Omega = 2\pi/T_0 = \Omega_0$
  - So, any pole of X(s) at s_α becomes periodic in $X^*(s)$ at $\alpha + jn\Omega_0$

Periodic Poles (and zeroes) of Starred Transform

- From previous slide, the starred transform of **_causal_** x(t) is $X^*(s)$
  - And $X^*(s)$ is periodic, as proven here:

$$X^*(s + jk2\pi/T_0) = \sum_{n=0}^{\infty} x(nT_0)e^{-n(s+jk2\pi/T_0)T_0} = \sum_{n=0}^{\infty} x[n]e^{-n\sigma T_0}e^{-jn(\Omega+k2\pi/T_0)T_0}$$

$$= \sum_{n=0}^{\infty} x[n]e^{-n\sigma T_0}e^{-jn\Omega T_0}e^{-jnk2\pi} = \sum_{n=-\infty}^{\infty} x[n]u[n]e^{-n\sigma T_0}e^{-jn\Omega T_0} = X^*(s)$$

> **Recall:**
> **Starred transform is restricted to causal signals**

- So, in the s-plane (assuming no aliasing):

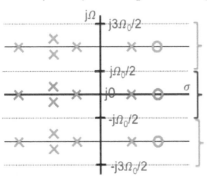

Periodic repetition

Primary strip

Periodic repetition

Correction to X*(s) for Discontinuities at t=0

- Note: certain forms of X*(s) **MUST** be corrected if there discontinuities
- Because inverse Laplace of X(s) converges to mean value of discontinuity
- Most commonly we are concerned with discontinuity at t=0
- Correction for a discontinuity at t=0, where **causal** x(t)=0 for t<0 :

$$x^*(t) = x(t) \sum_{n=-\infty}^{\infty} \delta(t - nT_0) = 0.5x(0)\delta(t) + L^{-1}\{X(s)\} \sum_{n=-\infty}^{\infty} \delta(t - nT_0)$$

$$= 0.5x(0)\delta(t) + L^{-1}\{X(s)\} \sum_{n=-\infty}^{\infty} \frac{e^{jn\Omega_0 t}}{T_0} \quad where \sum_{n=-\infty}^{\infty} \delta(t - nT_0) = \sum_{n=-\infty}^{\infty} \frac{e^{jn\Omega_0 t}}{T_0}$$

taking Laplace :

$$X^*(s) = 0.5x(0) + \frac{1}{T_0} \sum_{n=-\infty}^{\infty} X(s - jn\Omega_0) \quad where \quad e^{j\Omega_0 t} f(t) \Leftrightarrow F(s - j\Omega_0)$$

- So, **summarizing** for causal x(t) with discontinuity at t=0:

Periodic repetition

$$X^*(s) = \sum_{n=-\infty}^{\infty} x(nT_0)e^{-nsT_0} = X(z)\big|_{z=e^{sT_0}} = \boxed{0.5x(0)} + \frac{1}{T_0} \sum_{n=-\infty}^{\infty} X(s - jn\Omega_0)$$

These forms always OK   This form needs correction

Block Diagram for Starred Transform and ZOH

- Recall, from the model of an ideal sampler from a previous slide:

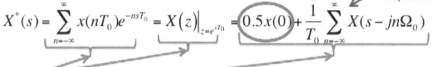

- After adding a plant function $G_P(s)$ to the output of the above system, and denoting the ZOH as $G_{H0}(s)$, the system block diagram becomes:

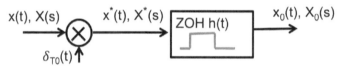

- However, a system such as the one above is most commonly represented as follows with the ZOH not explicitly drawn in the system, but absorbed into G(s):

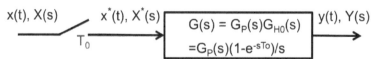

Summary: Block Diagram for Starred Transform and ZOH

- Summarizing from previous slide:

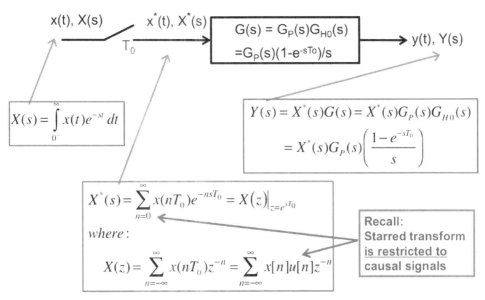

$$X(s) = \int_{0^-}^{\infty} x(t)e^{-st}\, dt$$

$$Y(s) = X^*(s)G(s) = X^*(s)G_P(s)G_{H0}(s)$$
$$= X^*(s)G_P(s)\left(\frac{1 - e^{-sT_0}}{s}\right)$$

$$X^*(s) = \sum_{n=0}^{\infty} x(nT_0)e^{-nsT_0} = X(z)\Big|_{z=e^{sT_0}}$$

where :

$$X(z) = \sum_{n=-\infty}^{\infty} x(nT_0)z^{-n} = \sum_{n=-\infty}^{\infty} x[n]u[n]z^{-n}$$

Recall:
Starred transform
is restricted to
causal signals

Spectra Through Block Diagram

Starred transform deals with causal

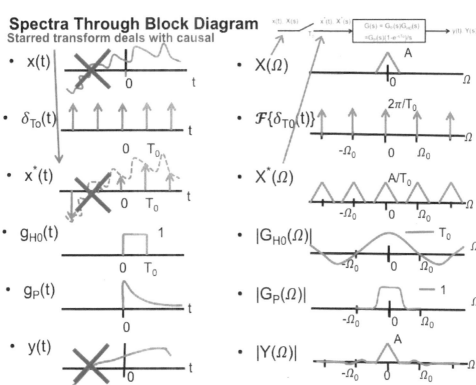

Summary: Spectra Through Block Diagram

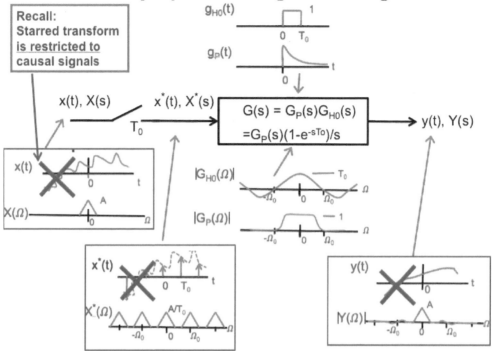

Example: Block Diagram Analysis With Sampler and ZOH

- From previous slide, let $x(t)=e^{-2t}u(t)$ and $T_0=0.1$ s, and $G_P(s)=1/s$,
  - Find $X(s)$, $X^*(s)$, and $Y(s)$:

x(t), X(s) $x^*(t)$, $X^*(s)$ $G(s) = G_P(s)G_{H0}(s)$ y(t), Y(s)

$=G_P(s)(1-e^{-sT_0})/s$

$$X(s) = \int_{0^-}^{\infty} x(t)e^{-st}\, dt = \int_{0^-}^{\infty} e^{-2t}e^{-st}\, dt = \int_{0^-}^{\infty} e^{-(s+2)t}\, dt = \frac{1}{s+2}$$

$$X(z) = \sum_{n=0}^{\infty} x(nT_0)z^{-n} = \sum_{n=-\infty}^{\infty} e^{-2nT_0}u[n]z^{-n} = \sum_{n=0}^{\infty}\left(e^{-2T_0}z^{-1}\right)^n = \frac{1}{1-e^{-2T_0}z^{-1}}; |z| > \left|e^{-2T_0}\right|$$

$$X^*(s) = \sum_{n=-\infty}^{\infty} x(nT_0)u(nT_0)e^{-nsT_0} = X(z)\Big|_{z=e^{sT_0}} = \frac{1}{1-e^{-2T_0}e^{-sT_0}} = \frac{1}{1-e^{-(s+2)T_0}}$$

$$Y(s) = X^*(s)G(s) = X^*(s)G_P(s)\left(\frac{1-e^{-sT_0}}{s}\right) = \left(\frac{1}{1-e^{-(s+2)T_0}}\right)\left(\frac{1}{s}\right)\left(\frac{1-e^{-sT_0}}{s}\right)$$

$$\text{with } T_0 = 0.1, \quad Y(s) = \frac{1-e^{-s/10}}{s^2\left(1-e^{-(s+2)/10}\right)}$$

5 OPEN-LOOP DIGITAL SYSTEMS

The lecture notes in this chapter present the design and analysis of open-loop digital systems.

Open-Loop Digital Systems

Starred Transform X*(s) and ZOH Output

- Recall:
  - Laplace transform of sampled signal x*(t) is the starred transform X*(s)

$$X^*(s) = \sum_{n=-\infty}^{\infty} x(nT_0)u(nT_0)e^{-nsT_0} = X(z)\Big|_{z=e^{sT_0}} = 0.5x(0^+) + \frac{1}{T_0}\sum_{n=-\infty}^{\infty} X(s-jn\Omega_0)$$

This form requires correction

- And including ZOH the combined output $X_0(t)$ is:

$$x_0(t) = x^*(t) * \Pi\left(\frac{t-T_0/2}{T_0}\right) \quad \Leftrightarrow \quad X_0(s) = X^*(s)\left(\frac{1-e^{-sT_0}}{s}\right)$$

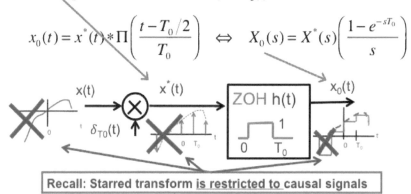

Recall: Starred transform is restricted to causal signals

70

Block Diagram Forms for Starred Transform and ZOH

- Also recall:
  - A system is most commonly represented as follows with the ZOH not explicitly drawn in the system, but absorbed into G(s)

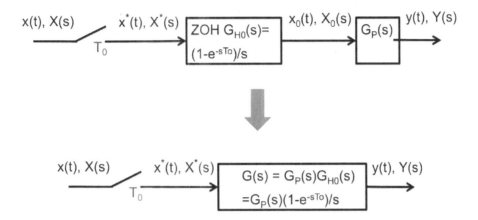

Digital Control System Model

- Similarly, the model of a digital control system now becomes:

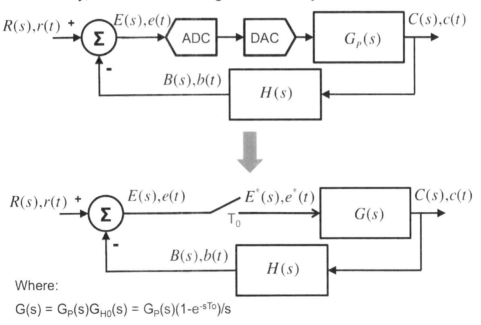

Where:

$G(s) = G_P(s)G_{H0}(s) = G_P(s)(1-e^{-sT_0})/s$

Properties of Starred Transform

Property 1: [e*(t) * g(t)]* = e*(t) * g*(t)

- Show that c*(t)=[e*(t) * g(t)]* = e*(t) * g*(t)

$$c(t) = \left(e(t) \sum_{n=-\infty}^{\infty} \delta(t - nT_0) \right) * g(t) = \int_{-\infty}^{\infty} g(\alpha) e(t - \alpha) \sum_{n=-\infty}^{\infty} \delta(t - \alpha - nT_0) d\alpha$$

$$= \sum_{n=-\infty}^{\infty} g(t - nT_0) e(t - t + nT_0) = \sum_{n=-\infty}^{\infty} e(nT_0) g(t - nT_0)$$

$$so: c^*(t) = c(t) \sum_{n=-\infty}^{\infty} \delta(t - nT_0) = \boxed{\sum_{m=-\infty}^{\infty} e(mT_0) g(t - mT_0) \sum_{n=-\infty}^{\infty} \delta(t - nT_0)}$$

$$and: e^*(t) * g^*(t) = g^*(t) * e^*(t) = \left(g(t) \sum_{m=-\infty}^{\infty} \delta(t - mT_0) \right) * \left(e(t) \sum_{n=-\infty}^{\infty} \delta(t - nT_0) \right)$$

Equal

$$= \int_{-\infty}^{\infty} \left(g(\alpha) \sum_{m=-\infty}^{\infty} \delta(\alpha - mT_0) \right) \left(e(t - \alpha) \sum_{n=-\infty}^{\infty} \delta(t - \alpha - nT_0) \right) d\alpha$$

$$= \sum_{n=-\infty}^{\infty} \sum_{m=-\infty}^{\infty} g(t - nT_0) e(t - t + nT_0) \delta(t - nT_0 - mT_0)$$

$$= \sum_{n=-\infty}^{\infty} \sum_{m=-\infty}^{\infty} e(nT_0) g(t - nT_0) \delta(t - (m + n)T_0) = \boxed{\sum_{n=-\infty}^{\infty} e(nT_0) g(t - nT_0) \sum_{\beta=-\infty}^{\infty} \delta(t - \beta T_0)}$$

- So: c*(t) = [e*(t) * g(t)]* = e*(t) * g*(t)

Property 2: [E*(s)G(s)]* = E*(s)G*(s)

- Show that [E*(s)G(s)]* = E*(s)G*(s)

  since:

  $$\left[e^*(t) * g(t) \right]^* = e^*(t) * g^*(t)$$

  then :

  By definition,

  $$L\left\{ \left[e^*(t) * g(t) \right]^* \right\} = L\left\{ e^*(t) * g^*(t) \right\}$$

  $L\{g^*(t)\}=G^*(s)$

  $$\left(L\left\{ \left[e^*(t) * g(t) \right] \right\} \right)^* = E^*(s)G^*(s)$$

  $$\boxed{\left[E^*(s)G(s) \right]^* = E^*(s)G^*(s)}$$

- So: [E*(s)G(s)]* = E*(s)G*(s)

Property 3: [X(s) (1-e⁻ˢᵀᵒ)]* = [X(s)]* (1-e⁻ˢᵀᵒ)

- Show that [X(s) (1-e⁻ˢᵀᵒ)]* = [X(s)]* (1-e⁻ˢᵀᵒ)

  $$\left[\left(X(s)\left(1 - e^{-sT_0}\right)\right) \right]^* = L\left\{ \left(x(t) - x(t - T_0) \right) \sum_{n=-\infty}^{\infty} \delta\left(t - nT_0 \right) \right\}$$

  $$= L\left\{ \left(x(t) \right) \sum_{n=-\infty}^{\infty} \delta\left(t - nT_0 \right) \right\} - L\left\{ \left(x(t - T_0) \right) \sum_{n=-\infty}^{\infty} \delta\left(t - nT_0 \right) \right\}$$

  $$= L\left\{ \left(x(t) \right) \sum_{n=-\infty}^{\infty} \delta\left(t - nT_0 \right) \right\} - L\left\{ \left(x(t - T_0) \right) \sum_{n=-\infty}^{\infty} \delta\left(t - nT_0 - T_0 \right) \right\}$$

  $$= L\left\{ \left(x(t) \right) \sum_{n=-\infty}^{\infty} \delta\left(t - nT_0 \right) \right\}\left(1 - e^{-sT_0}\right) = \left[X(s) \right]^* \left(1 - e^{-sT_0}\right)$$

  so :

  $$\boxed{\left[\left(X(s)\left(1 - e^{-sT_0}\right) \right) \right]^* = \left[X(s) \right]^* \left(1 - e^{-sT_0}\right)}$$

- So: [X(s) (1-e⁻ˢᵀᵒ)]* = [X(s)]* (1-e⁻ˢᵀᵒ)

Property 4: Z-Transform From Starred Transform

- The z-transform can be derived from the starred transform
- To see this, compare the two transforms

$$E(z) = Z\{e[n]\} = \sum_{n=0}^{\infty} e(nT_0)z^{-n} = \sum_{n=-\infty}^{\infty} e[n]u[n]z^{-n}$$

$$E^*(s) = L\{e^*(t)\} = \sum_{n=-\infty}^{\infty} e(nT_0)u(nT_0)e^{-snT_0} = \sum_{n=0}^{\infty} e[n]e^{-snT_0}$$

- So, by inspection:

$$E(z) = E^*(s)\Big|_{e^{sT_0}=z}$$

> Recall:
> Starred transform
> is restricted to
> causal signals

Generalizing Key Properties of Starred Transform

- **Restricted to causal signals only**

- For any X*(s) Y(s):

$$\left[X^*(s)Y(s)\right]^* = X^*(s)Y^*(s)$$

$$\left[\left(X(s)\left(1-e^{-sT_0}\right)\right)\right]^* = \left[X(s)\right]^*\left(1-e^{-sT_0}\right)$$

- Given X*(s) Y*(s), to find X(z) Y(z):

$$X(z)Y(z) = X^*(s)\Big|_{e^{sT_0}=z} Y^*(s)\Big|_{e^{sT_0}=z}$$

- Given X(z) Y(z), to find X*(s) Y*(s):

$$X^*(s)Y^*(s) = X(z)\Big|_{z=e^{sT_0}} Y(z)\Big|_{z=e^{sT_0}}$$

Open-Loop Transfer Function

Open-Loop Output C(s)

- Consider the following open-loop system

$$e(t), E(s) \quad\nearrow\quad e^*(t), E^*(s) \boxed{\begin{array}{c} G(s) = G_P(s)G_{H0}(s) \\ =G_P(s)(1\text{-}e^{\text{-sTo}})/s \end{array}} \quad c(t), C(s)$$

$$T_0$$

- Solve for C(s)

$$C(s) = E^*(s)G(s) = E^*(s)G_P(s)G_{H0}(s) = E^*(s)G_P(s)\left(\frac{1-e^{-sT_0}}{s}\right)$$

- Also, recall that starred transform E*(s) is periodic:
  - o Is periodic with period $2\pi/T_0 = \Omega_0$, so E*(s) = E*(s + jnΩ_0)
- But _C(s) is not necessarily periodic_
- Next:
  - o Find starred transform C*(s)

Open-Loop Output C*(s)

e(t), E(s) → e*(t), E*(s)

T_0

$G(s) = G_P(s)G_{H0}(s)$

$= G_P(s)(1-e^{-sT_0})/s$

c(t), C(s)

- From the previous slide:
$$C(s) = E^*(s)G(s)$$

Making use of "property 2"

- To find C*(s), take the starred transform C(s)
$$C^*(s) = \left[E^*(s)G(s)\right]^* = E^*(s)G^*(s)$$

where:
$$G(s) = G_P(s)G_{H0}(s) = G_P(s)\left(\frac{1-e^{-sT_0}}{s}\right)$$

- Including the original definition of starred transform for the above system:
$$C^*(s) = E^*(s)G^*(s) = \sum_{n=0}^{\infty} c(nT_0)e^{-snT_0}$$

Restricted to causal c(t)

- Where C*(s) is periodic, with period $2\pi/T_0 = \Omega_0$, so $C^*(s) = C^*(s + jn\Omega_0)$

Open-Loop Pulse Transfer Function G(z)

- The open-loop starred transform can now be used to derive the *open-loop pulse transfer function* G(z)

e(t), E(s) → e*(t), E*(s)

T_0

$G(s) = G_P(s)G_{H0}(s)$

$= G_P(s)(1-e^{-sT_0})/s$

c(t), C(s)

- From the open-loop system:
$$C^*(s) = \left[E^*(s)G(s)\right]^* = E^*(s)G^*(s) = \sum_{n=0}^{\infty} c(nT_0)e^{-nsT_0} = C(z)\Big|_{z=e^{sT_0}}$$

- Finding the z-transform from the starred transform:
$$E(z) = E^*(s)\Big|_{e^{sT_0}=z}$$

Restricted to causal E(z), G(z), C(z) and c(t)

so:
$$C(z) = C^*(s)\Big|_{e^{sT_0}=z} = E^*(s)\Big|_{e^{sT_0}=z} G^*(s)\Big|_{e^{sT_0}=z} = E(z)G(z)$$

- Where G(z)=C(z)/E(z) is the open-loop pulse transfer function

  o G(z) is the *transfer function from input to output at the sample instants*

Summary of Open-Loop Output C(s), C*(s), C(z)

$$e(t), E(s) \quad\quad e^*(t), E^*(s) \quad \boxed{\begin{array}{c} G(s) = G_P(s)G_{H0}(s) \\ =G_P(s)(1-e^{-sT_0})/s \end{array}} \quad c(t), C(s)$$

T_0

- Summarizing, for the open-loop system above

$$C(s) = E^*(s)G(s) = E^*(s)G_P(s)G_{H0}(s) = E^*(s)G_P(s)\left(\frac{1-e^{-sT_0}}{s}\right)$$

and :

$$C^*(s) = E^*(s)G^*(s) = \sum_{n=-\infty}^{\infty} c(nT_0)u(nT_0)e^{-nsT_0} = C(z)\Big|_{z=e^{sT_0}}$$

and :

$$C(z) = E(z)G(z) = C^*(s)\Big|_{e^{sT_0}=z}$$

> **Recall:**
> **Starred transform**
> **is restricted to**
> **causal signals**

- Where C*(s) is periodic, with period $2\pi/T_0 = \Omega_0$, so C*(s) = C*(s + jnΩ_0)

Example 1: Find G*(s) and G(z)

- Find the open-loop pulse transfer function G(z) and corresponding G*(s)
  - Let G(s) = $G_P(s)(1-e^{-sT_0})/s$, and $G_P(s)=1/(s+3)$

$$e(t), E(s) \quad\quad e^*(t), E^*(s) \quad \boxed{G(s) = G_P(s)(1-e^{-sT_0})/s} \quad c(t), C(s)$$

T_0

$$G(s) = G_P(s)\left(\frac{1-e^{-sT_0}}{s}\right) = \frac{1}{s+3}\left(\frac{1-e^{-sT_0}}{s}\right) = \frac{1}{s(s+3)}\left(1-e^{-sT_0}\right) = G'(s)\left(1-e^{-sT_0}\right)$$

$$g'(t) = L^{-1}\left\{\frac{1}{s(s+3)}\right\} = \frac{1}{3}\left(1-e^{-3t}\right)u(t);$$

$$G'(z) = \frac{1}{3}\sum_{n=0}^{\infty}\left(1-e^{-3nT_0}\right)z^{-n} = \frac{z/3}{z-1} - \frac{z/3}{z-e^{-3T_0}} = \frac{z\left(1-e^{-3T_0}\right)/3}{(z-1)\left(z-e^{-3T_0}\right)}; |z| > 1$$

$$G^*(s) = G'(z)\Big|_{z=e^{sT_0}}\left(1-e^{-sT_0}\right) = \frac{1}{3}\frac{z\left(1-e^{-3T_0}\right)\left(1-e^{-sT_0}\right)}{(z-1)\left(z-e^{-3T_0}\right)}\Big|_{z=e^{sT_0}} = \frac{1}{3}\left(\frac{1-e^{-3T_0}}{e^{sT_0}-e^{-3T_0}}\right)$$

$$G(z) = G^*(s)\Big|_{e^{sT_0}=z} = \frac{1}{3}\frac{1-e^{-3T_0}}{\left(z-e^{-3T_0}\right)}$$

Because
G*(s)=[G'(s) (1-e^{-sT0})]* = [G'(s)]* (1-e^{-sT0})

$$C(z) = E(z)G(z), \text{ so for } e[n]=u[n], C(z) = \frac{z}{z-1}G(z) = \frac{1}{3}\frac{z\left(1-e^{-3T_0}\right)}{(z-1)\left(z-e^{-3T_0}\right)},$$

$$and \ c[n] = g[n] = 0.333\left(1-e^{-3nT_0}\right)u[n] \text{ from the table} \longleftarrow \text{ Step response}$$

Example 1: dc Gain and $c(nT_0)$

- Find the dc gain
  - ○ Let $G(s) = G_P(s)(1-e^{-sT_0})/s$, and $G_P(s)=1/(s+3)$

$$e(t), E(s) \quad \diagup \quad e^*(t), E^*(s) \quad \boxed{G(s) = G_P(s)(1\text{-}e^{\text{-}sT_0})/s} \quad c(t), C(s)$$

$$T_0$$

- The dc gain from unit step, limit as $n \rightarrow \infty$:

$$\text{for } e[n]=u[n], \ C(z) = \frac{z}{z-1}G(z) = \frac{1}{3}\frac{z\left(1-e^{-3T_0}\right)}{(z-1)\left(z-e^{-3T_0}\right)},$$

$$\text{and } \lim_{n\to\infty} c[n] = \lim_{z\to 1}(1-z^{-1})\frac{1}{3}\frac{z\left(1-e^{-3T_0}\right)}{(z-1)\left(z-e^{-3T_0}\right)} = \frac{1}{3}$$

- The dc gain from $G_P(s)$ and $G(z)$:

$$G_P(s)\Big|_{s=0} = \frac{1}{s+3}\Big|_{s=0} = \frac{1}{3} \qquad G(z)\Big|_{z=1} = \frac{1}{3}\frac{1-e^{-3T_0}}{\left(z-e^{-3T_0}\right)}\Big|_{z=1} = \frac{1}{3}$$

- Output $c(nT_0)$ from $c[n]$:

$$c[n] = g[n] = 0.333\left(1-e^{-3nT_0}\right)u[n] \Rightarrow so: \ c(nT_0) = 0.333\left(1-e^{-3nT_0}\right)u(nT_0)$$

- Note: $c(nT_0)$ gives $c(t)$ only at sample points

Example 1: Find $C^*(s)$ and $C(s)$ for Unit Step Input

- Find the output $C(s)$ and corresponding $C^*(s)$ for $e(t)=u(t)$
  - ○ Let $G(s) = G_P(s)(1-e^{-sT_0})/s$, and $G_P(s)=1/(s+3)$

$$e(t), E(s) \quad \diagup \quad e^*(t), E^*(s) \quad \boxed{G(s) = G_P(s)(1\text{-}e^{\text{-}sT_0})/s} \quad c(t), C(s)$$

$$T_0$$

$$\text{for } e[n]=u[n], \ E(z) = \frac{z}{z-1} \quad and \quad E^*(s) = E(z)\Big|_{z=e^{sT_0}} = \frac{e^{sT_0}}{e^{sT_0}-1} = \frac{1}{1-e^{-sT_0}}$$

$$so: \ \boxed{C(s) = E^*(s)G(s) = E^*(s)G_P(s)\left(\frac{1-e^{-sT_0}}{s}\right) = \left(\frac{1}{1-e^{-sT_0}}\right)\left(\frac{1}{s+3}\right)\left(\frac{1-e^{-sT_0}}{s}\right) = \frac{1}{s(s+3)}}$$

from previous results: $G(z) = \frac{1}{3}\frac{1-e^{-3T_0}}{\left(z-e^{-3T_0}\right)}$

Step response

$$C(z) = E(z)G(z) \text{ and for } e[n]=u[n], \text{ and} \quad E(z) = \frac{z}{z-1}$$

$$C^*(s) = C(z)\Big|_{z=e^{sT_0}} = (E(z)G(z))\Big|_{z=e^{sT_0}} = \left(\frac{1}{3}\frac{z\left(1-e^{-3T_0}\right)}{(z-1)\left(z-e^{-3T_0}\right)}\right)\Bigg|_{z=e^{sT_0}}$$

$$\boxed{C^*(s) = \frac{1}{3}\frac{e^{sT_0}\left(1-e^{-3T_0}\right)}{\left(e^{sT_0}-1\right)\left(e^{sT_0}-e^{-3T_0}\right)} \quad \text{for } e[n]=u[n]}$$

Step response

Example 2: Find G*(s) G(z) for ZOH

- Find the open-loop pulse transfer function G(z) and corresponding G*(s)
  - Let G(s) = (1-e^{-sTo})/s, and $G_p(s)=1$

e(t), E(s)   e*(t), E*(s)   c(t), C(s)

T_0 →   G(s) = G_p(s)(1-e^{-sTo})/s →

$$G(s) = G_p(s)\left(\frac{1-e^{-sT_0}}{s}\right) = 1 \cdot \left(\frac{1-e^{-sT_0}}{s}\right) = \frac{1}{s}\left(1 - e^{-sT_0}\right) = G'(s)\left(1 - e^{-sT_0}\right)$$

$$g'(t) = L^{-1}\left\{\frac{1}{s}\right\} = u(t);$$

As before:

$$G^*(s) = [G'(s)\,(1\text{-}e^{\text{-sTo}})]^* = [G'(s)]^* \,(1\text{-}e^{\text{-sTo}})$$

$$G'(z) = \sum_{n=0}^{\infty} z^{-n} = \frac{z}{z-1}; |z| > 1$$

$$G^*(s) = G'(z)\Big|_{z=e^{sT_0}}\left(1 - e^{-sT_0}\right) = \frac{z\left(1-e^{-sT_0}\right)}{(z-1)}\Bigg|_{z=e^{sT_0}} = \frac{e^{sT_0}\left(1-e^{-sT_0}\right)}{\left(e^{sT_0}-1\right)} = 1$$

$$G(z) = G^*(s)\Big|_{e^{sT_0}=z} = 1$$

$C(z) = E(z)G(z) = E(z)$, so for e[n]=u[n], $c[n] = e[n] = u[n]$ ← Step response

- So, a ZOH by itself has pulse transfer function G(z)=1

- G(z) is the *transfer function from input to output at the sample instants*

Example 2: Find dc Gain of ZOH

- Find the open-loop pulse transfer function G(z) and corresponding G*(s)
  - Let G(s) = (1-e^{-sTo})/s, and $G_p(s)=1$

e(t), E(s)   e*(t), E*(s)   c(t), C(s)

T_0 →   G(s) = G_p(s)(1-e^{-sTo})/s →

- The dc gain from unit step, limit as n→∞ :

$$\text{for } e[n]=u[n],\ C(z) = \frac{z}{z-1}G(z) = \frac{z}{z-1},$$

$$and \lim_{n\to\infty} c[n] = \lim_{z\to 1}(1 - z^{-1})\frac{z}{z-1} = 1 \quad \leftarrow$$

- The dc gain from $G_p(s)$:

$$G_p(s)\Big|_{s=0} = 1 \quad \leftarrow$$

- The dc gain from G(z)

$$G(z)\Big|_{z=1} = 1\Big|_{z=1} = 1 \quad \leftarrow$$

- So, a ZOH by itself has dc gain of unity

Pulse Transfer Function

For

Various Open Loop Configurations

Pulse Transfer Function of Configuration 1

- Find the open-loop pulse transfer function G(z) for the following system:

$e_1(t), E_1$ (s) $\quad T_0 \quad$ $E_1^*(s)$ $\boxed{G_1(s)}$ $E_2(s)$ $\quad E_2^*(s)$ $\quad T_0 \quad$ $\boxed{G_2(s)}$ c(t), C(s)

- For the above open-loop system:

$$E_2^*(s) = \left[E_1^*(s)G_1(s)\right]^* = E_1^*(s)G_1^*(s)$$

$$C^*(s) = \left[E_2^*(s)G_2(s)\right]^* = E_2^*(s)G_2^*(s) = E_1^*(s)G_1^*(s)G_2^*(s)$$

$so :$

$$C(z) = C^*(s)\Big|_{e^{sT_0}=z} = E_1(z)G_1(z)G_2(z) = E_1(z)G(z)$$

- Where G(z)=C(z)/E₁(z)=G₁(z)G₂(z) is the open-loop pulse transfer function
  - G(z) is the *transfer function from input to output at the sample instants*

Pulse Transfer Function of Configuration 2

- Find the open-loop pulse transfer function G(z) for the following system:

$$e(t), E(s) \quad \xrightarrow{\quad} \quad e^*(t), E^*(s) \quad T_0 \quad \boxed{G_1(s)} \longrightarrow \boxed{G_2(s)} \quad c(t), C(s) \xrightarrow{\quad}$$

- For the above open-loop system:

$$C^*(s) = \left[E^*(s)G_1(s)G_2(s)\right]^* = E^*(s)\overline{G_1 G_2}^*(s)$$

$$C(z) = C^*(s)\Big|_{e^{sT_0}=z} = E(z)\overline{G_1 G_2}^*(z) = E(z)G(z) \qquad \begin{array}{l}\text{It is not possible}\\ \text{to split } G_1 \text{ and } G_2!\end{array}$$

overall pulse transfer function is: $G(z) = \overline{G_1 G_2}^*(z)$

- Where $\{G_1(s)G_2(s)\}^* \neq G_1^*(s)G_2^*(s)$

$$\overline{G_1 G_2}^*(s) = L\left\{\left(g_1(t) * g_2(t)\right) \sum_{n=-\infty}^{\infty} \delta(t - nT_0)\right\}$$

$$\overline{G_1 G_2}^*(s) \neq G_1^*(s)G_2^*(s)$$

- See next side for $\{G_1(s)G_2(s)\}^* \neq G_1^*(s)G_2^*(s)$

Example Showing $\{G_1(s)G_2(s)\}^* \neq G_1^*(s)G_2^*(s)$

- Example: choose G1(s)=G2(s)=1/s, so g1(t)=g2(t)=u(t)

$$e(t), E(s) \quad \xrightarrow{\quad} \quad e^*(t), E^*(s) \quad T_0 \quad \boxed{G_1(s)} \longrightarrow \boxed{G_2(s)} \quad c(t), C(s) \xrightarrow{\quad}$$

- For the above open-loop system:

$$\overline{G_1 G_2}^*(s) = L\left\{\left(g_1(t)*g_2(t)\right)\sum_{n=-\infty}^{\infty}\delta(t-nT_0)\right\} = L\left\{t\,u(t)\sum_{n=-\infty}^{\infty}\delta(t-nT_0)\right\} \qquad =0 \text{ at } n=0$$

$$= \overline{G_1 G_2}(z)\Big|_{z=e^{sT_0}} = \sum_{n=0}^{\infty} nT_0\, z^{-n}\Big|_{z=e^{sT_0}} = \frac{T_0 z}{(z-1)^2}\Big|_{z=e^{sT_0}} \quad ; \text{ and } \ \overline{g_1 g_2}[n] = T_0\, n\, u[n]$$

$$G_1^*(s)G_2^*(s) = \{G_1(z)G_2(z)\}\Big|_{z=e^{sT_0}} = \left\{\left(\sum_{n=0}^{\infty}1z^{-n}\right)\left(\sum_{n=0}^{\infty}1z^{-n}\right)\right\}\Big|_{z=e^{sT_0}} = \frac{z^2}{(z-1)^2}\Big|_{z=e^{sT_0}} \qquad \text{Not equal}$$

$$and \quad g_1[n]*g_2[n] = T_0(n+1)u[n+1] = T_0(n+1)u[n] \qquad =T_0 \text{ at } n=0$$

$$so: \ \overline{G_1 G_2}^*(s) \neq G_1^*(s)G_2^*(s) \ \text{ since } \ \frac{T_0 z}{(z-1)^2}\Big|_{z=e^{sT_0}} \neq \frac{z^2}{(z-1)^2}\Big|_{z=e^{sT_0}} \quad and \ \overline{G_1 G_2}(z) \neq G_1(z)G_2(z)$$

- So, in general, $\{G_1(s)G_2(s)\}^* \neq G_1^*(s)G_2^*(s)$

Note {1,1,1..}*{1,1,1,...}
={1,2,3...} ≠ {0,1,2,3...}

Pulse Transfer Function of Configuration 3

- Find the open-loop pulse transfer function G(z) for the following system:

r(t), R(s) $\rightarrow$ | $G_1(s)$ | $\xrightarrow{E(s)}$ $\xrightarrow{T_0}$ $\xrightarrow{E^*(s)}$ | $G_2(s)$ | $\xrightarrow{c(t),\ C(s)}$

- For the above open-loop system:

$$E^*(s) = \left[R(s)G_1(s)\right]^* = \overline{RG_1}^*(s)$$

It is not possible to split R and G_1!

$$where: \quad \overline{RG_1}^*(s) = L\left\{\left(r(t) * g_1(t)\right) \sum_{n=-\infty}^{\infty} \delta(t - nT_0)\right\}$$

$$C^*(s) = \left[E^*(s)G_2(s)\right]^* = E^*(s)G_2^*(s) = \overline{RG_1}^*(s)\, G_2^*(s)$$

$$C(z) = C^*(s)\big|_{e^{sT_0}=z} = G_2(z)\,\overline{RG_1}\,(z)$$

$$note: \quad \overline{RG_1}^*(s) = \overline{G_1R}^*(s) \quad and \quad \overline{RG_1}(z) = \overline{G_1R}(z)$$

- Above, it is not possible to find the open-loop pulse transfer function G(z)=C(z)/R(z)
- Because R(z) cannot be de-embedded to form C(z) = R(z)G(z)

Pulse Transfer Function
With
Digital Filters Included

Block Diagram Including Digital Filter D(z)

- Now, add digital filtering D(z) to the system
    - As before, system is most commonly represented as follows with the ZOH not explicitly drawn in the system, but absorbed into $G_C(s)$

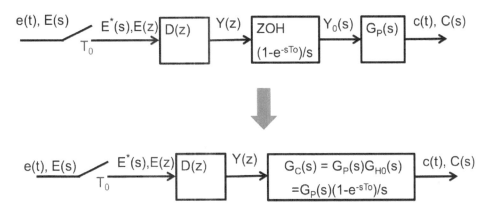

Pulse Transfer Function of Configuration 4

- The open-loop pulse transfer function G(z) for other configurations

- For the above open-loop system:

$$Y^*(s) = Y(z)\big|_{z=e^{sT_0}} = E(z)D(z)\big|_{z=e^{sT_0}} = E^*(s)D^*(s)$$

$$C^*(s) = \left[Y^*(s)G_C(s)\right]^* = Y^*(s)G_C^*(s) = E^*(s)D^*(s)G_C^*(s)$$

so :

$$\boxed{\begin{array}{l} C(z) = C^*(s)\big|_{e^{sT_0}=z} = E(z)D(z)G_C(z) = E(z)G(z) \\[2mm] and \quad G(z) = D(z)G_C(z) \end{array}}$$

- Where G(z) =C(z)/E(z)=D(z)G_C(z) is the open-loop pulse transfer function
    - G(z) is the *transfer function from input to output at the sample instants*

Example 3: Find G*(s) and G(z)

- Find the open-loop pulse transfer function G(z) and corresponding G*(s)
  - Let $G_C(s) = G_P(s)(1-e^{-sT_0})/s$, and $G_P(s)=1/(s+5)$, and $D(z)=1+z^{-1}$

$$e(t), E(s) \quad\nearrow\quad E^*(s), E(z) \quad \boxed{D(z)} \quad Y^*(s), Y(z) \quad \boxed{\begin{array}{c} G_C(s) = G_P(s)G_{HO}(s) \\ = G_P(s)(1-e^{-sT_0})/s \end{array}} \quad c(t), C(s)$$

$$T_0$$

$$G_C(s) = \frac{1}{s(s+5)}\left(1-e^{-sT_0}\right) = G'(s)\left(1-e^{-sT_0}\right); \quad so \ g'(t) = \frac{1}{5}\left(1-e^{-5t}\right)u(t);$$

$$G'(z) = \frac{1}{5}\sum_{n=0}^{\infty}\left(1-e^{-5nT_0}\right)z^{-n} = \frac{z/5}{z-1} - \frac{z/5}{z-e^{-5T_0}} = \frac{z\left(1-e^{-5T_0}\right)/5}{(z-1)\left(z-e^{-5T_0}\right)}; |z|>1$$

$$G_C^*(s) = G'(z)\Big|_{z=e^{sT_0}}\left(1-e^{-sT_0}\right) = \frac{1}{5}\frac{z\left(1-e^{-5T_0}\right)\left(1-e^{-sT_0}\right)}{(z-1)\left(z-e^{-5T_0}\right)}\Bigg|_{z=e^{sT_0}} = \frac{1}{5}\frac{1-e^{-5T_0}}{\left(e^{sT_0}-e^{-5T_0}\right)}$$

$$G_C(z) = G_C^*(s)\Big|_{e^{sT_0}=z} = \frac{1}{5}\frac{1-e^{-5T_0}}{\left(z-e^{-5T_0}\right)}$$

Note: dc
gain = 2/5

$$C(z) = E(z)D(z)G_C(z) = E(z)\left(\frac{z+1}{z}\right)\left(\frac{1}{5}\frac{1-e^{-5T_0}}{\left(z-e^{-5T_0}\right)}\right) = E(z)\frac{1}{5}\frac{(z+1)\left(1-e^{-5T_0}\right)}{z\left(z-e^{-5T_0}\right)}$$

$$and \quad G(z) = D(z)G_C(z) \quad and \quad C^*(s) = C(z)\Big|_{z=e^{sT_0}} = E^*(s)\frac{1}{5}\frac{\left(e^{sT_0}+1\right)\left(1-e^{-5T_0}\right)}{e^{sT_0}\left(e^{sT_0}-e^{-5T_0}\right)}$$

Fractional Time Delays

and

Modified Z-Transform

Modified Z-Transform

- Define modified z-transform for <u>fractional time delay</u> τ

$$let: \quad \tau = (k + \Delta)T_0 \quad where: \quad 0 < \Delta < 1$$

and define the modified z-transform as:

$$Z\{y(t - T_0\Delta)\} = \sum_{n=-\infty}^{\infty} y(nT_0 - T_0\Delta)z^{-n} = Y(z, 1 - \Delta); \quad for \ 0 < \Delta < 1$$

- Then, the z-transform for any general time delay becomes

$$Z\{x(t - \tau)\} = z^{-k}Z\{x(t - T_0\Delta)\} = z^{-k}X(z, 1 - \Delta)$$

- These are tabulated in appendix as X(z,m) with m=1-Δ
- The modified z-transform can then be used to find the corresponding starred transform for fractional time delay τ

$$X^*(s) = e^{-skT_0}L\left\{(x(t - T_0\Delta))\sum_{n=-\infty}^{\infty}\delta(t - nT_0)\right\}$$

$$= X(z)\Big|_{z=e^{sT_0}} = z^{-k}X(z, 1 - \Delta)\Big|_{z=e^{sT_0}}$$

Table of Some Modified z-transforms

Mod. z-transform, X(z,m)	z-transform	Time Func	Laplace Transform
	1	$\delta(t)$	1
$\dfrac{1}{z-1}$	$\dfrac{z}{z-1}; \quad |z| > 0$	$u(t)$	$1/s$
$\dfrac{mT_0}{z-1} + \dfrac{T_0}{(z-1)^2}$	$\dfrac{zT_0}{(z-1)^2}; \quad |z| > 1$	$tu(t)$	$1/s^2$
$\dfrac{e^{-amT_0}}{z-e^{-aT_0}}$	$\dfrac{z}{z-e^{-aT_0}}; \quad |z| > e^{-aT_0}$	$e^{-at}u(t)$	$\dfrac{1}{s+a}$
$\dfrac{T_0e^{-amT_0}\left[e^{-aT_0} + m\left(z - e^{-aT_0}\right)\right]}{\left(z - e^{-aT_0}\right)^2}$	$\dfrac{T_0ze^{-aT_0}}{\left(z-e^{-aT_0}\right)^2}; \quad |z| > e^{-aT_0}$	$te^{-at}u(t)$	$\dfrac{1}{(s+a)^2}$

Pulse Transfer Function with Time Delay

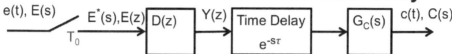

- For the above open-loop system, signal processing includes time delay τ
- May arise from computation time, and is modeled in Laplace as $e^{-s\tau}$
- Integer delays of "k" clock periods are easily merged into D(z) as $z^{-k}D(z)$
- Fractional delays require the modified z-transform

$$let: \quad \tau = (k+\Delta)T_0$$

$$C(z) = E(z)D(z)z^{-k}G_C(z,1-\Delta) = E(z)G(z)$$

$$so: \quad G(z) = z^{-k}D(z)G_C(z,1-\Delta)$$

$$and: \quad C^*(s) = C(z)\big|_{z=e^{sT_0}} = \left(E(z)D(z)z^{-k}G_C(z,1-\Delta)\right)\Big|_{z=e^{sT_0}}$$

- Where $G(z)=z^{-k}D(z)G_C(z,1-\Delta)$ is the open-loop pulse transfer function
  - $G(z)$ is the _transfer function from input to output at the sample instants_

6 CLOSED-LOOP DIGITAL SYSTEMS

The lecture notes in this chapter present the design and analysis of closed-loop digital systems.

Closed-Loop Digital Control Systems

Digital Control System Model

- Recall: closed-loop digital control system:

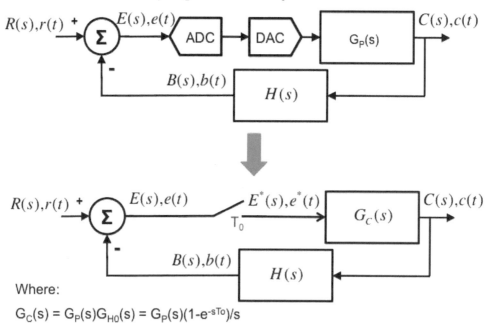

Where:

$G_C(s) = G_P(s)G_{H0}(s) = G_P(s)(1-e^{-sT_0})/s$

Digital Control System Model

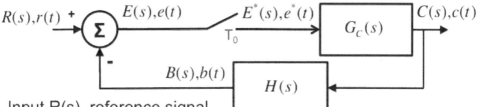

- Input R(s), reference signal
- Output C(s), controlled signal
- Feedback B(s), feedback signal (back)
- Error E(s), error signal (ideally zero if C(s)=R(s))
- $G_C(s)$ is forward path continuous-time transfer (or plant) function
  - Includes ZOH: $G_C(s) = G_P(s)G_{H0}(s) = G_P(s)(1-e^{-sT_0})/s$
- H(s) is feedback transfer function
- $G_{OL}(s)=G_C(s)H(s)$ is open-loop transfer function
- $G_{CL}(z)=C(z)/R(z)$ is (closed-loop) pulse transfer function
- $G_{CL}(s)=C(s)/R(s)$ is closed-loop transfer function

Analysis of Closed-Loop Systems

- Feedback systems
  - Potential for instability
  - Analysis required to assure stability (topic in later lectures)
  - Analysis provides insight for compensation to stabilize
- Closed loop analysis for mixed discrete/continuous-time systems
  - Similar to continuous-time systems analysis
  - However:
    - Requires starred transform and z-transform
    - Replace ADC/DAC and D(z) by samplers
    - Absorb ZOH into $G_C(s)$
- Demonstrate technique by example
  - Examine a number of configurations/topologies

Digital Control System: Configuration 1

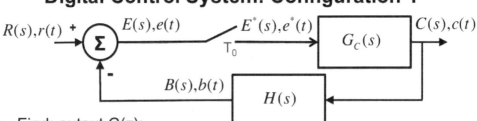

- Find: output C(z):

$$E(s) = R(s) - B(s) = R(s) - C(s)H(s) = R(s) - E^*(s)G_C(s)H(s)$$

$$solving: \quad E^*(s) = \frac{R^*(s)}{1 + \overline{G_C H}^*(s)}$$

Take starred transform of both sides of eqns.

$$then: \quad C(s) = E^*(s)G_C(s)$$

$$so: \quad C^*(s) = \left[E^*(s)G_C(s)\right]^* = E^*(s)G_C^*(s) = \frac{G_C^*(s)}{1 + \overline{G_C H}^*(s)}R^*(s)$$

$$and \quad C(z) = \frac{G_C(z)}{1 + \overline{G_C H}(z)}R(z) = G_{CL}(z)R(z); \quad where: G_{CL}(z) = \frac{G_C(z)}{1 + \overline{G_C H}(z)}$$

- $G_{CL}(z)$ is closed-loop pulse transfer function

Digital Control System: Configuration 2

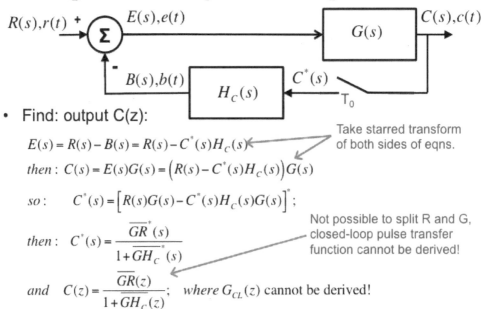

- Find: output C(z):

$$E(s) = R(s) - B(s) = R(s) - C^*(s)H_C(s)$$

Take starred transform of both sides of eqns.

$$then: \quad C(s) = E(s)G(s) = \left(R(s) - C^*(s)H_C(s)\right)G(s)$$

$$so: \quad C^*(s) = \left[R(s)G(s) - C^*(s)H_C(s)G(s)\right]^*;$$

Not possible to split R and G, closed-loop pulse transfer function cannot be derived!

$$then: \quad C^*(s) = \frac{\overline{GR}^*(s)}{1 + \overline{GH_C}^*(s)}$$

$$and \quad C(z) = \frac{\overline{GR}(z)}{1 + \overline{GH_C}(z)}; \quad where \ G_{CL}(z) \ cannot \ be \ derived!$$

- $G_{CL}(z) = C(z)/R(z)$ cannot be derived!

Digital Control System: Configuration 3

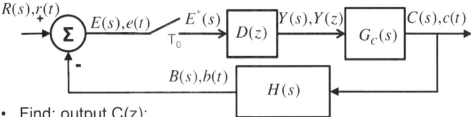

- Find: output C(z):

$$E(s) = R(s) - B(s) = R(s) - C(s)H(s) = R(s) - E^*(s)D(z)\big|_{z=e^{sT_0}} G_C(s)H(s)$$

$solving:$ $\quad E^*(s) = \dfrac{R^*(s)}{1 + D^*(s)\overline{G_C H}^{\,*}(s)}$ $\quad where: D^*(s) = D(z)\big|_{z=e^{sT_0}}$ $\quad$ Take starred transform of both sides of eqns.

$then:$ $\quad C(s) = E^*(s)D^*(s)G_C(s)$

$so:$ $\quad C^*(s) = \left[E^*(s)D^*(s)G_C(s)\right]^* = E^*(s)D^*(s)G_C^*(s) = \dfrac{D^*(s)G_C^*(s)}{1 + D^*(s)\overline{G_C H}^{\,*}(s)} R^*(s)$

$and \quad C(z) = C^*(s)\big|_{e^{sT_0}=z} = \dfrac{D(z)G_C(z)}{1 + D(z)\overline{G_C H}(z)} R(z) = G_{CL}(z)R(z);$

$where: G_{CL}(z) = \dfrac{D(z)G_C(z)}{1 + D(z)\overline{G_C H}(z)}$

- $G_{CL}(z)$ is closed-loop pulse transfer function

Digital Control System: Configuration 4

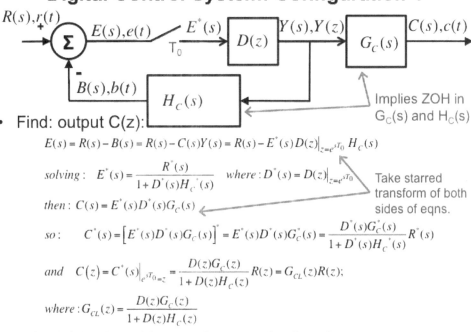

Implies ZOH in $G_C(s)$ and $H_C(s)$

- Find: output C(z):

$$E(s) = R(s) - B(s) = R(s) - C(s)Y(s) = R(s) - E^*(s)D(z)\big|_{z=e^{sT_0}} H_C(s)$$

$solving:$ $\quad E^*(s) = \dfrac{R^*(s)}{1 + D^*(s)H_C^*(s)}$ $\quad where: D^*(s) = D(z)\big|_{z=e^{sT_0}}$ $\quad$ Take starred transform of both sides of eqns.

$then:$ $\quad C(s) = E^*(s)D^*(s)G_C(s)$

$so:$ $\quad C^*(s) = \left[E^*(s)D^*(s)G_C(s)\right]^* = E^*(s)D^*(s)G_C^*(s) = \dfrac{D^*(s)G_C^*(s)}{1 + D^*(s)H_C^*(s)} R^*(s)$

$and \quad C(z) = C^*(s)\big|_{e^{sT_0}=z} = \dfrac{D(z)G_C(z)}{1 + D(z)H_C(z)} R(z) = G_{CL}(z)R(z);$

$where: G_{CL}(z) = \dfrac{D(z)G_C(z)}{1 + D(z)H_C(z)}$

- $G_{CL}(z)$ is closed-loop pulse transfer function

Digital Control System: Configuration 5

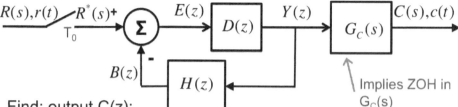

- Find: output C(z):

$$E(z) = R(z) - B(z) = R(z) - E(z)D(z)H(z)$$

$$solving: \quad E(z) = \frac{R(z)}{1 + D(z)H(z)} \quad and: E^*(s) = E(z)\big|_{z=e^{sT_0}} = \frac{R^*(s)}{1 + D^*(s)H^*(s)}$$

$$then: \quad C(s) = G_C(s)Y^*(s) = G_C(s)Y(z)\big|_{z=e^{sT_0}} = G_C(s)\big(E(z)D(z)\big)\big|_{z=e^{sT_0}} = G_C(s)E^*(s)D^*(s)$$

$$so: \quad C^*(s) = \left[G_C(s)E^*(s)D^*(s)\right]^* = G_C^*(s)E^*(s)D^*(s) = \frac{D^*(s)G_C^*(s)}{1 + D^*(s)H^*(s)}R^*(s)$$

$$and \quad C(z) = C^*(s)\big|_{e^{sT_0}=z} = \frac{D(z)G_C(z)}{1 + D(z)H(z)}R(z) = G_{CL}(z)R(z);$$

$$where: G_{CL}(z) = \frac{D(z)G_C(z)}{1 + D(z)H(z)}$$

- $G_{CL}(z)$ is closed-loop pulse transfer function

Importance of Starred Transform in Analysis

- Could not solve any of these problems without starred transform
- Output of every closed-loop configuration above required starred transform

$$configuration\ 1: \quad C(s) = E^*(s)G_C(s)$$

$$configuration\ 2: \quad C(s) = \big(R(s) - C^*(s)H_C(s)\big)G(s)$$

$$configuration\ 3: \quad C(s) = E^*(s)D^*(s)G_C(s)$$

$$configuration\ 4: \quad C(s) = E^*(s)D^*(s)G_C(s)$$

$$configuration\ 5: \quad C(s) = E^*(s)D^*(s)G_C(s)$$

- All open-loop configurations last week required starred transform

$$open\text{-}loop\ configuration\ 1: \quad C(s) = E_1^*(s)G_1^*(s)G_2(s)$$

$$open\text{-}loop\ configuration\ 2: \quad C(s) = E^*(s)G_1(s)G_2(s)$$

$$open\text{-}loop\ configuration\ 3: \quad C(s) = E^*(s)G_2(s)$$

$$open\text{-}loop\ configuration\ 4: \quad C(s) = E^*(s)D^*(s)G_C(s)$$

- Hence, the **central importance of starred transform** in analysis of mixed analog/digital circuits and systems

Digital Control System: Configuration 6

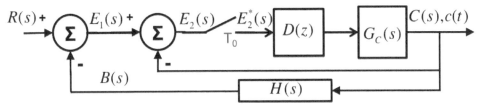

- Find: output C(z):

$$E_1(s) = R(s) - B(s) = R(s) - C(s)H(s) = R(s) - E_2^*(s)D^*(s)G_C(s)H(s)$$

$$E_2(s) = E_1(s) - C(s) = E_1(s) - E_2^*(s)D^*(s)G_C(s) \qquad where: D^*(s) = D(z)|_{z=e^{sT_0}}$$

$$so: \quad E_2^*(s) = E_1^*(s) - E_2^*(s)D^*(s)G_C^*(s) \quad or: E_1^*(s) = E_2^*(s)\left(1 + D^*(s)G_C^*(s)\right)$$

$$and: \quad E_1^*(s) = R^*(s) - E_2^*(s)D^*(s)\overline{G_C H}^*(s)$$

$$so: E_2^*(s) = R^*(s) / \left(1 + D^*(s)G_C^*(s) + D^*(s)\overline{G_C H}^*(s)\right)$$

$$so: \quad C^*(s) = \left[E_2^*(s)D^*(s)G_C(s)\right]^* = E_2^*(s)D^*(s)G_C^*(s) = \frac{D^*(s)G_C^*(s)}{1 + D^*(s)G_C^*(s) + D^*(s)\overline{G_C H}^*(s)}R^*(s)$$

$$and \quad C(z) = C^*(s)|_{e^{sT_0} = z} = \frac{D(z)G_C(z)}{1 + D(z)G_C(z) + D(z)\overline{G_C H}(z)}R(z) = G_{CL}(z)R(z);$$

- $G_{CL}(z)$ is closed-loop pulse transfer function

Digital Control System: Configuration 7 (delay)

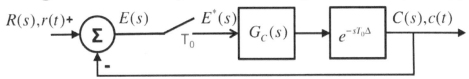

- Find: output C(z):

$$E(s) = R(s) - B(s) = R(s) - C(s) = R(s) - E^*(s)G_C(s)e^{-sT_0\Delta}$$

$$E^*(s) = R^*(s) - E^*(s)G_C(z, 1-\Delta)|_{z=e^{sT_0}}$$

$$solving: \quad E^*(s) = \frac{R^*(s)}{1 + G_C(z, 1-\Delta)|_{z=e^{sT_0}}} \qquad \text{Modified z-transform to absorb delay}$$

$$then: \quad C(s) = E^*(s)G_C(s)e^{-sT_0\Delta}$$

$$C^*(s) = \left[E^*(s)G_C(s)e^{-sT_0\Delta}\right]^* = E^*(s)G_C(z, 1-\Delta)|_{z=e^{sT_0}} = \frac{G_C(z, 1-\Delta)|_{z=e^{sT_0}}}{1 + G_C(z, 1-\Delta)|_{z=e^{sT_0}}}R^*(s)$$

$$and \quad C(z) = \frac{G_C(z, 1-\Delta)}{1 + G_C(z, 1-\Delta)}R(z) = G_{CL}(z)R(z); \quad where: G_{CL}(z) = \frac{G_C(z, 1-\Delta)}{1 + G_C(z, 1-\Delta)}$$

- $G_{CL}(z)$ is closed-loop pulse transfer function

Generalized Analysis Methods

Analysis Method 1: Block Diagram Direct Method

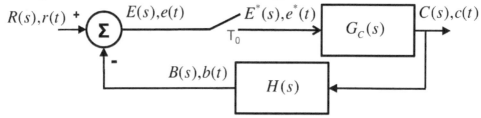

- A simple method to analyze system directly from block diagram:

1) express system output in terms of sampler outputs and system input and output:

$$C(s) = E^*(s)G_c(s)$$

2) express sampler inputs in terms of sampler outputs and system input and output:

$$E(s) = R(s) - B(s) = R(s) - E^*(s)G_c(s)H(s)$$

3) take starred transforms and solve by substituting into equations:

$$E^*(s) = R^*(s) - E^*(s)\overline{G_cH}^*(s); \text{ so } \quad E^*(s) = \frac{R^*(s)}{1 + \overline{G_cH}^*(s)}$$

$$C^*(s) = \left[E^*(s)G_c(s)\right]^* = E^*(s)G_c^*(s) = \frac{R^*(s)G_c^*(s)}{1 + \overline{G_cH}^*(s)} \quad \text{as before!}$$

Analysis Method 2: Flow Graph Method

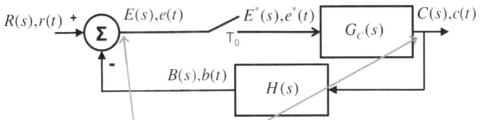

- Convert block diagram into graph:
  - Breaking samplers into nodes E(s) and inputs E*(s)
  - Each breakout/feedback point becomes a node
  - Each adder/subtractor output becomes a node

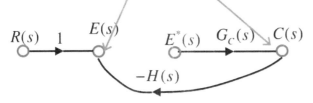

- Obtain equations from graph (see next slide)

Analysis Method 2, cont'd.: Equations from Flow Graph

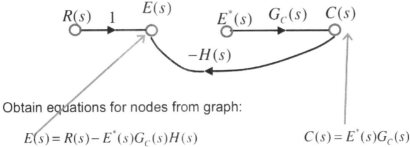

- Obtain equations for nodes from graph:

$$E(s) = R(s) - E^*(s)G_C(s)H(s) \qquad\qquad C(s) = E^*(s)G_C(s)$$

- Take starred transform of equations:

$$E^*(s) = \left[R(s) - E^*(s)G_C(s)H(s)\right]^* = R^*(s) - E^*(s)\overline{G_C H}^*(s)$$

$$C^*(s) = \left[E^*(s)G_C(s)\right]^* = E^*(s)G_C^*(s)$$

- Solve equations:
  - By subsitution (as before)
  - By Cramer's rule
  - By Mason's gain formula
  - Mathcad symbolic solve block

Equation Solution Methods

Solution Method 1: Solve by Substitution

- The basic approach in most of the proceeding examples
- Solve equations by:
  - o Taking starred transforms where needed
  - o Substituting to eliminate unknowns (E1*, E2*, etc.)

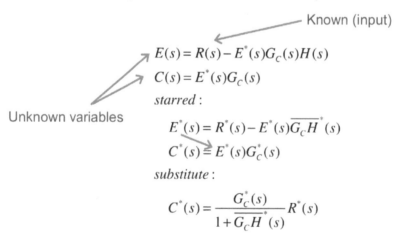

Known (input)

$$E(s) = R(s) - E^*(s)G_C(s)H(s)$$

$$C(s) = E^*(s)G_C(s)$$

Unknown variables

starred :

$$E^*(s) = R^*(s) - E^*(s)\overline{G_C H}^*(s)$$

$$C^*(s) = E^*(s)G_C^*(s)$$

substitute :

$$C^*(s) = \frac{G_C^*(s)}{1 + \overline{G_C H}^*(s)} R^*(s)$$

Solution Method 2: Solve by Cramer's Rule

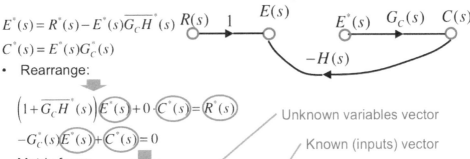

$$E^*(s) = R^*(s) - E^*(s)\overline{G_cH}^*(s)$$
$$C^*(s) = E^*(s)G_c^*(s)$$

- Rearrange:

$$\left(1 + \overline{G_cH}^*(s)\right)E^*(s) + 0 \cdot C^*(s) = R^*(s)$$

$$-G_c^*(s)E^*(s) + C^*(s) = 0$$

- Matrix form:

$$\begin{bmatrix} 1 + \overline{G_cH}^*(s) & 0 \\ -G_c^*(s) & 1 \end{bmatrix} \begin{bmatrix} E^*(s) \\ C^*(s) \end{bmatrix} = \begin{bmatrix} R^*(s) \\ 0 \end{bmatrix}$$

Unknown variables vector
Known (inputs) vector

- Cramer's rule:

$$E^*(s) = \frac{\begin{vmatrix} R^*(s) & 0 \\ 0 & 1 \end{vmatrix}}{\begin{vmatrix} 1 + \overline{G_cH}^*(s) & 0 \\ -G_c^*(s) & 1 \end{vmatrix}} = \frac{R^*(s)}{1 + \overline{G_cH}^*(s)}; \quad C^*(s) = \frac{\begin{vmatrix} 1 + \overline{G_cH}^*(s) & R^*(s) \\ -G_c^*(s) & 0 \end{vmatrix}}{\begin{vmatrix} 1 + \overline{G_cH}^*(s) & 0 \\ -G_c^*(s) & 1 \end{vmatrix}} = \frac{R^*(s)G_c^*(s)}{1 + \overline{G_cH}^*(s)}$$

Solution Method 3: Solve by Mason's Formula

$$E^*(s) = R^*(s) - E^*(s)\overline{G_cH}^*(s)$$
$$C^*(s) = E^*(s)G_c^*(s)$$
$$C(s) = E^*(s)G_c(s)$$

- First, draw a new graph based on starred equations, inputs, and outputs
- Then apply Mason's gain formula to find transfer function
- Where:
  - G_k is gain from R* to C* along forward path k
  - Δ_{All} is one minus sum of all individual loop gains (for this example only)
  - Δ_k excluding all loops touching the G_k path

$$\frac{C^*(s)}{R^*(s)} = \frac{1}{\Delta_{All}} \sum_{k=1}^{fwd paths} G_k \Delta_k = \frac{1}{1 + \overline{G_cH}^*(s)} \left(G_c^*(s)\right)(1) = \frac{G_c^*(s)}{1 + \overline{G_cH}^*(s)}$$

- Mason's rule: see textbook appendix for details

Digital Control System: Example 1

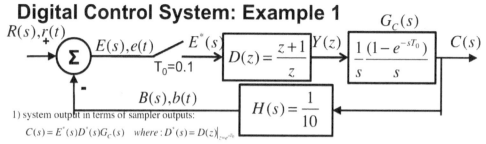

1) system output in terms of sampler outputs:

$$C(s) = E^*(s)D^*(s)G_C(s) \quad where: D^*(s) = D(z)\big|_{z=e^{sT_0}}$$

2) express sampler inputs in terms of sampler outputs and system input and output:

$$E(s) = R(s) - C(s)H(s) = R(s) - E^*(s)D^*(s)G_C(s)H(s)$$

3) take starred transforms and solve:

$$E^*(s) = R^*(s) - E^*(s)D^*(s)\overline{G_C H}^*(s); \; so \quad E^*(s) = \frac{R^*(s)}{1 + D^*(s)\overline{G_C H}^*(s)}$$

$$G_C(s) = \frac{1 - e^{-sT_0}}{s^2} = G'(s)\left(1 - e^{-sT_0}\right);$$

$$\Rightarrow g'(t) = t\, u(t);$$

$$G'(z) = \frac{zT_0}{(z-1)^2}; \quad |z| > 1$$

$$C^*(s) = E^*(s)D^*(s)G_C^*(s) = \frac{R^*(s)D^*(s)G_C^*(s)}{1 + D^*(s)\overline{G_C H}^*(s)} = \frac{\frac{z+1}{z}\frac{zT_0}{(z-1)^2}\Big|_{z=e^{sT_0}}(1-e^{-sT_0})}{1 + \frac{z+1}{z}\frac{zT_0}{10(z-1)^2}\Big|_{z=e^{sT_0}}(1-e^{-sT_0})} R^*(s)$$

dc gain=10

$$and \quad C(z) = \frac{D(z)G_C(z)}{1 + D(z)\overline{G_C H}(z)} R(z) = \frac{\left(\frac{z+1}{z}\right)\left(\frac{zT_0}{(z-1)^2}\right)\left(\frac{z-1}{z}\right)}{1 + \left(\frac{z+1}{z}\right)\left(\frac{zT_0}{10(z-1)^2}\right)\left(\frac{z-1}{z}\right)} R(z) = \frac{(z+1)T_0}{z(z-1) + (z+1)T_0/10} R(z);$$

Digital Control System: Example 1

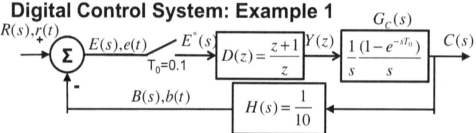

- Finally, we would like to solve for C(s)
- Returning to our equations:

$$C(s) = E^*(s)D^*(s)G_C(s) \quad where: D^*(s) = D(z)\big|_{z=e^{sT_0}}$$

$$E(s) = R(s) - E^*(s)D^*(s)G_C(s)H(s) \quad or \quad E^*(s) = \frac{R^*(s)}{1 + D^*(s)\overline{G_C H}^*(s)} \quad \text{Cannot}$$
$$\text{eliminate E(s)}$$

$$so: \quad C(s) = \frac{R(s) - E(s)}{D^*(s)G_C(s)H(s)} D^*(s)G_C(s) = \frac{R(s) - E(s)}{H(s)}$$

- So, cannot solve for C(s) in terms of R(s)
- In this system, can solve for C(s) in terms of R*(s) only:

$$C(s) = \frac{R^*(s)D^*(s)G_C(s)}{1 + D^*(s)\overline{G_C H}^*(s)} \quad and \quad c(t) = L^{-1}\{C(s)\}$$

- More about this topic next week

7 CLOSED-LOOP TIME RESPONSE

The lecture notes in this chapter discuss the time-domain response of closed-loop digital systems.

Closed-Loop Time Response

Closed-Loop System Time Response

- Last time:
  - o Began analysis of feedback systems
  - o Derived closed-loop pulse transfer function for many systems: $G_{CL}(z)=C(z)/R(z)$
  - o Note: some systems, could not derive $C(z)/R(z)$
- Next:
  - o Focus on finding $C(s)$ and $c(t)$ for many examples
  - o After finding $C(s)$, $c(t)= L^{-1}\{C(s)\}$
  - o Would like to find some $G_{CL}(s)=C(s)/R(s)$
  - o But in general will have to find $C(s)$ as function of $R^*(s)$
  - o Nevertheless, result is sufficient to find $C(s)$ and $c(t)$

Digital Control System Model

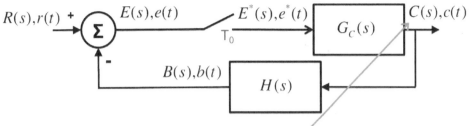

- Last time:
  - Looked at a number of configurations
  - Solved for $G_{CL}(z) = C(z)/R(z)$ closed-loop pulse transfer function
- Next:
  - Focus on finding C(s)
  - Then, $c(t) = L^{-1}\{C(s)\}$

Finding c(t) from C(s) and R*(s)

Step 1

Find C(s)

Method illustrated by examples

Digital Control System: Configuration 1

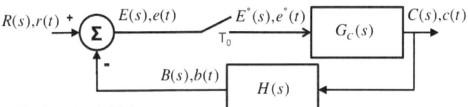

- Find: output C(s):

$$C(z) = \frac{G_C(z)}{1+G_C H(z)} R(z)$$

$$E(s) = R(s) - E^*(s)G_C(s)H(s) \quad or: \quad E^*(s) = \frac{R^*(s)}{1+G_C H^*(s)}$$

$$C(s) = E^*(s)G_C(s)$$

$$so: \quad C(s) = \left(R^*(s) / \left(1+\overline{G_C H}^*(s) \right) \right) G_C(s) \quad or \quad \boxed{C(s) = \frac{R^*(s)G_C(s)}{1+\overline{G_C H}^*(s)}}$$

- So, cannot solve for C(s) in terms of R(s)
- In this system, can solve for C(s) in terms of R*(s) only:
- And, c(t)= L⁻¹{C(s)}

Digital Control System: Configuration 2

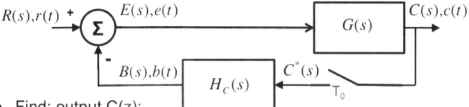

- Find: output C(z):

$$C(z) = \frac{\overline{GR}(z)}{1 + \overline{GH}_c(z)}$$

$E(s) = R(s) - B(s) = R(s) - C^*(s)H_c(s)$ Not possible to remove C*!

$C(s) = E(s)G(s)$

$then: \ C(s) = E(s)G(s) = \left(R(s) - C^*(s)H_c(s) \right)G(s)$

- C(s) cannot be derived!

Digital Control System: Configuration 3

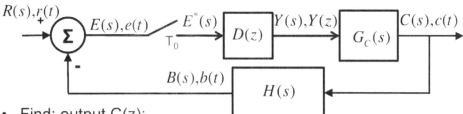

- Find: output C(z):

$$C(z) = \frac{D(z)G_c(z)}{1 + D(z)\overline{G_cH}(z)}R(z)$$

$E(s) = R(s) - E^*(s)D^*(s)G_c(s)H(s) \quad or: E^*(s) = \dfrac{R^*(s)}{1 + D^*(s)\overline{G_cH}^*(s)}$ Cannot eliminate E(s)

$C(s) = E^*(s)D^*(s)G_c(s)$

$so: \ C(s) = \dfrac{R(s) - E(s)}{H(s)} \qquad or: \ \boxed{C(s) = \dfrac{R^*(s)D^*(s)G_c(s)}{1 + D^*(s)\overline{G_cH}^*(s)}}$

- So, cannot solve for C(s) in terms of R(s)
- In this system, can solve for C(s) in terms of R*(s) only:
- And, c(t)= L⁻¹{C(s)}
-

Digital Control System: Configuration 4

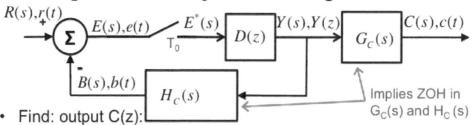

- Find: output C(z):

$$C(z) = \frac{D(z)G_C(z)}{1 + D(z)H_C(z)} R(z)$$

$$E(s) = R(s) - E^*(s)D^*(s)H_C(s) \quad or: E^*(s) = \frac{R^*(s)}{1 + D^*(s)H_C^*(s)}$$

$$C(s) = E^*(s)D^*(s)G_C(s)$$

Cannot eliminate E(s)

$$so: \quad C(s) = \frac{(R(s) - E(s))G_C(s)}{H_C(s)} \quad or: \quad C(s) = \frac{R^*(s)D^*(s)G_C(s)}{1 + D^*(s)H_C^*(s)}$$

- So, cannot solve for C(s) in terms of R(s)
- In this system, can solve for C(s) in terms of R*(s) only:
- And, c(t)= L⁻¹{C(s)}
-

Digital Control System: Configuration 5

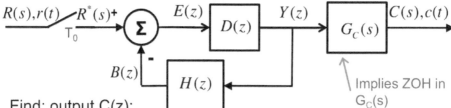

- Find: output C(z):

$$C(z) = \frac{D(z)G_C(z)}{1 + D(z)H(z)} R(z)$$

$$E(z) = R(z) - E(z)D(z)H(z) \quad or: E^*(s) = \frac{R^*(s)}{1 + D^*(s)H^*(s)}$$

$$C(s) = E^*(s)D^*(s)G_C(s)$$

$$so: \quad C(s) = \frac{R^*(s)D^*(s)G_C(s)}{1 + D^*(s)H^*(s)}$$

- So, cannot solve for C(s) in terms of R(s)
- In this system, can solve for C(s) in terms of R*(s) only:
- And, c(t)= L⁻¹{C(s)}
-

Digital Control System: Configuration 6

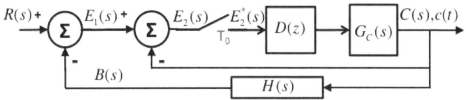

- Find: output C(z):

$$C(z) = \frac{D(z)G_C(z)}{1 + D(z)G_C(z) + D(z)\overline{G_C H}(z)} R(z)$$

$$E_1(s) = R(s) - C(s)H(s) = R(s) - E_2^*(s)D^*(s)G_C(s)H(s)$$

$$E_2(s) = E_1(s) - E_2^*(s)D^*(s)G_C(s) \quad or: E_2^*(s) = \frac{R^*(s)}{1 + D^*(s)G_C^*(s) + D^*(s)\overline{G_C H}^*(s)}$$

$$C(s) = E_2^*(s)D^*(s)G_C(s)$$

so: $\boxed{C(s) = \dfrac{R^*(s)D^*(s)G_C(s)}{1 + D^*(s)G_C^*(s) + D^*(s)\overline{G_C H}^*(s)}}$

- So, cannot solve for C(s) in terms of R(s)
- In this system, can solve for C(s) in terms of R*(s) only:
- And, c(t)= L⁻¹{C(s)}
-

Digital Control System: Configuration 7 (delay)

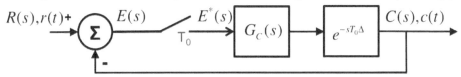

- Find: output C(z):

Modified z-transform to absorb delay

$$C(z) = \frac{G_C(z, 1-\Delta)}{1 + G_C(z, 1-\Delta)} R(z)$$

$$E(s) = R(s) - E^*(s)G_C(s)e^{-sT_0\Delta} \quad or: E^*(s) = \frac{R^*(s)}{1 + G_C(z, 1-\Delta)\big|_{z=e^{sT_0}}}$$ Cannot eliminate E(s)

$$C(s) = E^*(s)G_C(s)e^{-sT_0\Delta}$$

so: $C(s) = R(s) - E(s) \quad or: \boxed{C(s) = \dfrac{R^*(s)G_C(s)e^{-sT_0\Delta}}{1 + G_C(z, 1-\Delta)\big|_{z=e^{sT_0}}}}$

- So, cannot solve for C(s) in terms of R(s)
- In this system, can solve for C(s) in terms of R*(s) only:
- And, c(t)= L⁻¹{C(s)}
-

Finding c(t) from C(s) and R*(s)

Step 2

Find c(t)

Method illustrated by examples

Summary: Finding c(t) from C(s)

- Previous examples:
  - o Derived C(s)

- Next:
  - o After finding C(s),
    - – c(t)= $L^{-1}\{C(s)\}$
  - o Approach to finding $L^{-1}\{C(s)\}$
    - – Generally will require breaking C(s) into:
      - > z-domain portion
      - > and s-domain portion

General Approach: Finding c(t) from C(s)

- General approach to finding output c(t) from C(s):
  - First, express R*(s) as r[n] z^{-n} and combine other z-components in $C(s)=G_{CL}(s)R^*(s)$ where $G_{CL}(s)=G_{CLs}(s)$ $G_{CLz}(z)|_{z=exp(sTo)}$
  - Express combined as z-transform sum with terms of form e^{-nsTo}
  - Then use Laplace delay property ← z-components

$$C(s) = G_{CL}(s)R^{'}(s) = G_{CLs}(s)\left(G_{CLz}(z)R(z)\right)\Big|_{z=e^{sT_0}} = G_{CLs}(s)\overline{G_{CLz}R(z)}\Big|_{z=e^{sT_0}}$$

← s-components

$$where: \overline{G_{CLz}R(z)}\Big|_{z=e^{sT_0}} = \sum_{n=0}^{\infty}\overline{g_{CLz}r(nT_0)}z^{-n}\Big|_{z=e^{sT_0}} = \sum_{n=0}^{\infty}\overline{g_{CLz}r(nT_0)}e^{-nsT_0}$$

Laplace delay property

$$so: $$

$$C(s) = G_{CLs}(s)\sum_{n=0}^{\infty}\overline{g_{CLz}r(nT_0)}e^{-nsT_0} = \sum_{n=0}^{\infty}\overline{g_{CLz}r(nT_0)}\left(G_{CLs}(s)e^{-nsT_0}\right)$$

$$and:$$

$$\boxed{c(t) = \sum_{n=0}^{\infty}\overline{g_{CLz}r(nT_0)}L^{-1}\left\{G_{CLs}(s)e^{-nsT_0}\right\} = \sum_{n=0}^{\infty}\overline{g_{CLz}r[n]}g_{CLs}(t-nT_0)}$$

- So, c(t) is the sum of delayed impulse responses weighted by r[n]

Recall: **Starred transform is restricted to causal signals**

Finding c(t) from C(s) and R*(s)

The full two-step method is best illustrated by example

Example 1: First review finding C(z) (Configuration 1)

$R(s), r(t)$ + Σ — $E(s), e(t)$ — $E^*(s)$ — $T_0 = 0.1$ — $G_C(s) = 30 \dfrac{(1 - e^{-sT_0})}{s}$ — $C(s), c(t)$

$B(s), b(t)$ — $H(s) = \dfrac{1}{10}$

- Find: output C(z):

$$solving: \quad E^*(s) = \frac{R^*(s)}{1 + \overline{G_c H}^*(s)} = \frac{R^*(s)}{1 + \frac{3z(1 - e^{-sT_0})}{(z-1)}\Big|_{z=e^{sT_0}}}$$

$$G_c(s)H(s) = 3\frac{(1 - e^{-sT_0})}{s} = G'(s)\left(1 - e^{-sT_0}\right);$$
$$\Rightarrow g'(t) = 3u(t); \quad \Rightarrow G'(z) = 3z/(z-1)$$
$$\overline{G_c H}^*(s) = \frac{3z(1 - e^{-sT_0})}{(z-1)}\Big|_{z=e^{sT_0}} = 3; \; |z| > 1$$

$$so: C^*(s) = E^*(s)G_C^*(s) = \frac{\frac{30z(1 - e^{-sT_0})}{(z-1)}\Big|_{z=e^{sT_0}} R^*(s)}{1 + \frac{3z(1 - e^{-sT_0})}{(z-1)}\Big|_{z=e^{sT_0}}} = \frac{30z(1 - e^{-sT_0})\Big|_{z=e^{sT_0}} R^*(s)}{(z-1) + 3z(1 - e^{-sT_0})\Big|_{z=e^{sT_0}}}$$

show this equals 3

$$and \quad C(z) = \frac{G_C(z)}{1 + G_c H(z)} R(z) = \frac{30z(1 - z^{-1})}{(z-1) + 3z(1 - z^{-1})} R(z) = \frac{30}{4} R(z)$$

dc gain=30/4

- $G_{CL}(z) = 30/4$ is closed-loop pulse transfer function

Example 1: Step 1 Find C(s) (Configuration 1)

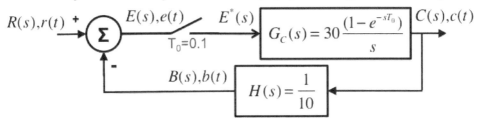

- Step 1 to find c(t):
  o Find output C(s):

$$E(s)=R(s)-B(s)=R(s)-C(s)H(s)=R(s)-E^*(s)G_C(s)H(s)$$

$$solving: \quad E^*(s)=\frac{R^*(s)}{1+\overline{G_CH}^*(s)}$$ ← from last time

$$then: \quad C(s)=E^*(s)G_C(s)$$

- Next step:

Example 1: Step 2 Find c(t) (Configuration 1)

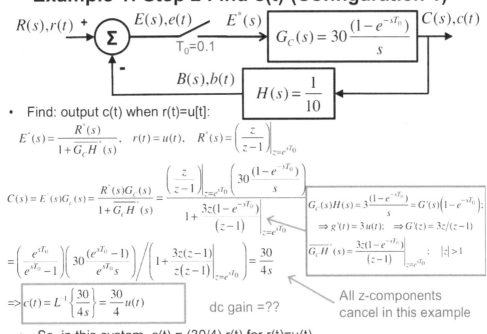

- Find: output c(t) when r(t)=u[t]:

$$E^*(s)=\frac{R^*(s)}{1+\overline{G_CH}^*(s)}, \quad r(t)=u(t), \quad R^*(s)=\left(\frac{z}{z-1}\right)\Big|_{z=e^{sT_0}}$$

$$C(s)=E^*(s)G_c(s)=\frac{R^*(s)G_C(s)}{1+\overline{G_CH}^*(s)}=\frac{\left(\frac{z}{z-1}\right)\Big|_{z=e^{sT_0}}30\frac{(1-e^{-sT_0})}{s}}{1+\frac{3z(1-e^{-sT_0})}{(z-1)}\Big|_{z=e^{sT_0}}}$$

$$=\left(\frac{e^{sT_0}}{e^{sT_0}-1}\right)\left(30\frac{(e^{sT_0}-1)}{e^{sT_0}s}\right)\Big/\left(1+\frac{3z(z-1)}{z(z-1)}\Big|_{z=e^{sT_0}}\right)=\frac{30}{4s}$$

$$\Rightarrow \boxed{c(t)=L^{-1}\left\{\frac{30}{4s}\right\}=\frac{30}{4}u(t)}$$ dc gain =??

$$G_C(s)H(s)=3\frac{(1-e^{-sT_0})}{s}=G'(s)\left(1-e^{-sT_0}\right);$$
$$\Rightarrow g'(t)=3u(t); \quad \Rightarrow G'(z)=3z/(z-1)$$
$$\overline{G_CH}^*(s)=\frac{3z(1-e^{-sT_0})}{(z-1)}\Big|_{z=e^{sT_0}}; \quad |z|>1$$

All z-components cancel in this example

- So, in this system, c(t) = (30/4) r(t) for r(t)=u(t)

Example 2: First review finding C(z) (Configuration 1)

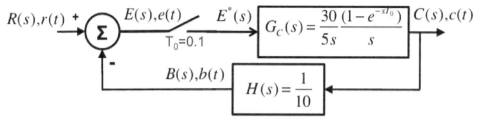

- Find: output C(z):

$$solving: \quad E^*(s) = \frac{R^*(s)}{1+\overline{G_C H}^*(s)} = \frac{R^*(s)}{1+\frac{3T_0}{5(z-1)}\Big|_{z=e^{sT_0}}}$$

$$G_C(s)H(s) = \frac{3}{5s}\frac{(1-e^{-sT_0})}{s} = G'(s)(1-e^{-sT_0});$$
$$\Rightarrow g'(t) = 0.6t\,u(t); \Rightarrow G'(z) = 0.6T_0 z/(z-1)^2$$
$$\overline{G_C H}^*(s) = \frac{3T_0 z(1-e^{-sT_0})}{5(z-1)^2}\Big|_{z=e^{sT_0}} = \frac{3T_0}{5(z-1)}\Big|_{z=e^{sT_0}}$$

$$so: C^*(s) = E^*(s)G_C^*(s) = \frac{\frac{30T_0}{5(z-1)}\Big|_{z=e^{sT_0}} R^*(s)}{1+\frac{3T_0}{5(z-1)}\Big|_{z=e^{sT_0}}} = \frac{30T_0 R^*(s)}{5(z-1)\Big|_{z=e^{sT_0}} + 3T_0}$$

$$and \quad C(z) = \frac{G_C(z)}{1+G_C H(z)}R(z) = \frac{30T_0}{5(z-1)+3T_0}R(z) = G_{CL}(z)R(z) \quad dc\ gain=10$$

- $G_{CL}(z)$ is closed-loop pulse transfer function

Example 2: Step 1 Find C(s) (Configuration 1)

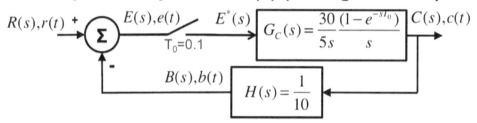

- Step 1 to find c(t):
  - Find output C(s):

$$E(s) = R(s) - B(s) = R(s) - C(s)H(s) = R(s) - E^*(s)G_C(s)H(s)$$

$$solving: \quad E^*(s) = \frac{R^*(s)}{1+\overline{G_C H}^*(s)}$$
$$then: \quad C(s) = E^*(s)G_C(s)$$

from last time

- Next step:

Example 2: Step 2 Find c(t) (Configuration 1)

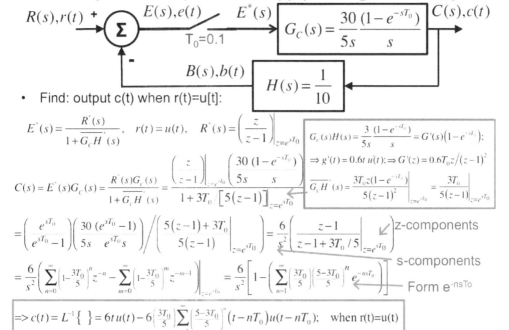

- Find: output c(t) when r(t)=u[t]:

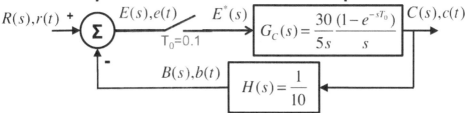

Example 2: Note on z and s components

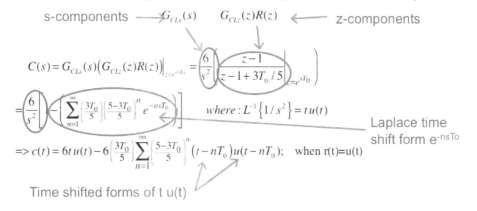

- Notes on decomposition into z-component and s-component

Example 2: Simulation (Configuration 1)

$$C(s) = \frac{6}{s^2}\left[1 - \left(\sum_{n=1}^{\infty}\left(\tfrac{3T_0}{5}\right)\left(\tfrac{5-3T_0}{5}\right)^n e^{-nsT_0}\right)\right] => c(t) = 6t\,u(t) - 6\left(\tfrac{3T_0}{5}\right)\sum_{n=1}^{\infty}\left(\tfrac{5-3T_0}{5}\right)^n\left(t - nT_0\right)u(t - nT_0)$$

$$\text{cdirClass}(t, T_0) := \frac{6}{1}\left[t - \left(\frac{3T_0}{5}\right)\cdot\sum_{n=0}^{200}\left[\left(\frac{5 - 3T_0}{5}\right)^n\cdot(t - n\cdot T_0 - T_0)\cdot u(t - n\cdot T_0 - T_0)\right]\right]$$

$T0 = 0.1$

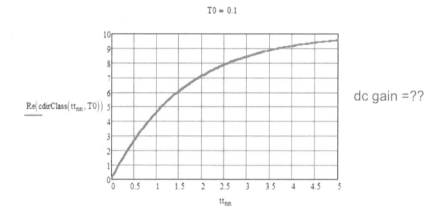

dc gain =??

8 CLOSED-LOOP STABILITY

This chapter covers a variety of methods for design and analysis of the closed-loop stability of digital control systems.

Closed-Loop Stability

Closed-Loop System Stability

- Last time:
    - ○ Time-domain analysis of feedback systems
    - ○ Focus on finding C(s) and c(t) for many examples

- Next:
    - ○ Focus on stability of closed-loop discrete-time systems
    - ○ Main approach: poles and zeroes of $G_{CL}(z)$
        - – But beware of pole/zero cancellation and hidden modes
    - ○ Other methods in text
        - – Routh-Hurwitz
        - – Jury
    - ○ Other common approaches can falsely indicate stable
        - – See papers by Stephen D. Stearns for more info

System Characteristic Equation

- For closed-loop systems we found $G_{CL}(z)=C(z)/R(z)=N(z)/D(z)$
  - Where $N(z)$ and $D(z)$ are the numerator and denominator polynomials
- Using partial fraction expansion (assume first-order poles):

$$G_{CL}(z)=\frac{N(z)}{D(z)}=\frac{z\,Q(z)}{D(z)}=\frac{z\,Q(z)}{(z-p_1)(z-p_2)\circ\circ\circ(z-p_i)}=z\sum_{\alpha=1}^{i}\frac{c_\alpha}{z-p_\alpha}$$

$$where:\ a^n u[n]\Leftrightarrow\frac{z}{z-a}:\ \ |z|\triangleright|a|$$

$$So:\ g_{CL}(z)[n]=Z^{-1}\{G_{CL}(z)\}=Z^{-1}\left\{\sum_{\alpha=1}^{i}\frac{zc_\alpha}{z-p_\alpha}\right\}=\sum_{\alpha=1}^{i}c_\alpha\left(p_\alpha\right)^n u[n]$$

- Clearly, the poles of $G_{CL}(z)$ (roots of $D(z)$) determine system behavior
  - Poles inside the unit circle are well-behaved
  - Poles outside unit circle diverge
  - Equation $D(z) = 0$ is called the system characteristic equation
- For configuration 1, as an example:

More precisely, we later also include denominator of $G_C(z)$

$$G_{CL}(z)=\frac{C(z)}{R(z)}=\frac{G_C(z)}{1+\overline{G_C H}(z)}$$

so characteristic equation is: $1+\overline{G_C H}(z)=0$

Digital Control System: Configuration 1

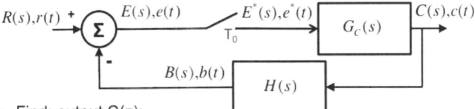

- Find: output C(z):

$$E(s)=R(s)-B(s)=R(s)-C(s)H(s)=R(s)-E^*(s)G_C(s)H(s)$$

$$solving:\ E^*(s)=\frac{R^*(s)}{1+\overline{G_C H}^*(s)}$$

Take starred transform of both sides of eqns.

$$then:\ C(s)=E^*(s)G_C(s)$$

$$so:\ \ \ C^*(s)=\left[E^*(s)G_C(s)\right]^*=E^*(s)G_C^*(s)=\frac{G_C^*(s)}{1+\overline{G_C H}^*(s)}R^*(s)$$

$$and\ \ \ G_{CL}(z)=\frac{C(z)}{R(z)}=\frac{G_C(z)}{1+\overline{G_C H}(z)}$$

Other Closed-Loop Configurations

- Each closed-loop configuration has a different characteristic equation
- For each characteristic equation, the zeroes set system poles

More precisely, we later also include denominator of numerators

configuration 1: $G_{CL}(z) = G_C(z)/\left(1 + \overline{G_C H}(z)\right)$

configuration 2: $C(z) = \overline{GR}(z)/\left(1 + \overline{GH_C}(z)\right)$

configuration 3: $G_{CL}(z) = D(z)G_C(z)/\left(1 + D(z)\overline{G_C H}(z)\right)$

configuration 4: $G_{CL}(z) = D(z)G_C(z)/\left(1 + D(z)H_C(z)\right)$

Characteristic equation

configuration 5: $G_{CL}(z) = D(z)G_C(z)/\left(1 + D(z)H(z)\right)$

configuration 6: $G_{CL}(z) = D(z)G_C(z)/\left(1 + D(z)G_C(z) + D(z)\overline{G_C H}(z)\right)$

configuration 7: $G_{CL}(z) = G_C(z, 1-\Delta)/\left(1 + G_C(z, 1-\Delta)\right)$

- Note that configuration 2 it was not possible to find $G_{CL}(z)$
- See text for dealing with such cases

Note on Characteristic Equation

- Consider

$$G_{CL}(z) = \frac{C(z)}{R(z)} = \frac{G_C(z)}{1 + G_C H(z)} = \frac{G_{CN}(z)/G_{CD}(z)}{1 + G_C H_N(z)/G_C H_D(z)}$$

More precisely, characteristic equation must include denominator of the numerator

so characteristic equation is NOT this:

$$1 + \overline{G_C H}(z) = 0 \quad or \quad \overline{G_C H_N}(z) + \overline{G_C H_D}(z) = 0$$

- More carefully, the denominator includes :

$$G_{CL}(z) = \frac{C(z)}{R(z)} = \frac{G_C(z)}{1 + G_C H(z)} = \frac{G_{CN}(z)/G_{CD}(z)}{1 + G_C H_N(z)/G_C H_D(z)} = \frac{G_{CN}(z)}{G_{CD}(z) + G_{CD}(z)G_C H_N(z)/G_C H_D(z)}$$

so characteristic equation is:

$$G_{CD}(z) + G_{CD}(z)\overline{G_C H_N}(z)/\overline{G_C H_D}(z) = 0 \quad or: \quad G_{CD}(z)\left(1 + \overline{G_C H_N}(z)/\overline{G_C H_D}(z)\right) = 0$$

if G and H are separable, it becomes easier:

$$\frac{G_N(z)/G_D(z)}{1 + G_N(z)H_N(z)/\left(G_D(z)H_D(z)\right)}$$

$$\Rightarrow G_D(z) + G_N(z)H_N(z)/H_D(z) = 0 \quad \Rightarrow \quad G_D(z)H_D(z) + G_N(z)H_N(z) = 0$$

also, note that at a pole in Gc(z):

$$G_{CL}(z) = \frac{C(z)}{R(z)} \approx \frac{\infty}{1 + \infty H(z)} = \frac{1}{H(z)}$$

But beware pole-zero cancellation, see comments later

Starred Transform Characteristic Equation

- For pulse transfer function

$$G_{CL}(z) = \frac{C(z)}{R(z)} = \frac{G_C(z)}{1 + G_C H(z)} = \frac{G_{CN}(z)/G_{CD}(z)}{1 + G_C H_N(z)/G_C H_D(z)}$$

so characteristic equation is NOT this:

$$1 + \overline{G_C H(z)} = 0 \quad or \quad \overline{G_C H_N(z)} + \overline{G_C H_D(z)} = 0$$

More precisely, characteristic equation must include denominator of the numerator

- Starred transform form of characteristic equation is

$$G_{CL}^*(s) = \frac{C^*(s)}{R^*(s)} = \frac{G_C^*(s)}{1 + \overline{G_C H}^*(s)} = \frac{\overline{G_{CN}^*(s)/G_{CD}^*(s)}}{1 + \overline{G_C H_N}^*(s)/\overline{G_C H_D}^*(s)}$$

$$= \frac{G_{CN}^*(s)}{G_{CD}^*(s) + G_{CD}^*(s)\overline{G_C H_N}^*(s)/\overline{G_C H_D}^*(s)}$$

so characteristic equation is:

$$G_{CD}^*(s)\left(1 + \overline{G_C H_N}^*(s)/\overline{G_C H_D}^*(s)\right) = 0$$

- Stability is determined in Laplace plane
- All poles must be in left-half s-plane to be stable

Comment on Pole-Zero Cancellation

- For pulse transfer function

$$G_{CL}(z) = \frac{C(z)}{R(z)} = \frac{G_C(z)}{1 + G_C H(z)} = \frac{G_{CN}(z)/G_{CD}(z)}{1 + G_C H_N(z)/G_C H_D(z)}$$

- If we had pole zero cancellation at $z = \beta$

$$G_{CL}(z) = \frac{C(z)}{R(z)} = \frac{G_N(z)D_N(z)/\left(G_D(z)D_D(z)\right)}{1 + G_N(z)D_N(z)/\left(G_D(z)D_D(z)\right)} = \frac{(z-\beta)G_N'(z)D_N(z)/\left((z-\beta)G_D(z)D_D'(z)\right)}{1 + (z-\beta)G_N'(z)D_N(z)/\left((z-\beta)G_D(z)D_D'(z)\right)} =$$

$$= \frac{(z-\beta)G_N'(z)D_N(z)}{\left((z-\beta)G_D(z)D_D'(z)\right) + (z-\beta)G_N'(z)D_N(z)} = \frac{(z-\beta)G_N'(z)D_N(z)}{(z-\beta)\left(G_D(z)D_D'(z) + G_N'(z)D_N(z)\right)}$$

- So the zero still appears in the denominator
- Use this form/method to avoid problems
- **Do not "cancel" the pole/zero pair at any earlier stages**

Summary: Stability Analysis

- For a stable closed-loop system in z-domain:
  - All poles of $G_{CL}(z)$ must be inside of unit circle
  - For configuration 1:

$$G_{CL}(z) = \frac{C(z)}{R(z)} = \frac{G_C(z)}{1 + G_C H(z)} = \frac{G_{CN}(z)/G_{CD}(z)}{1 + G_C H_N(z)/G_C H_D(z)}$$

- For a stable closed-loop system in s-domain:
  - All poles of $G^*_{CL}(s)$ must be inside left-half s-plane
  - For configuration 1:

$$G^*_{CL}(s) = \frac{C^*(s)}{R^*(s)} = \frac{G^*_C(s)}{1 + \overline{G_C H}^*(s)} = \frac{G^*_{CN}(s)/G^*_{CD}(s)}{1 + \overline{G_C H_N}^*(s)/\overline{G_C H_D}^*(s)}$$

- Other configurations have different $G^*_{CL}(s)$ and $G_{CL}(z)$

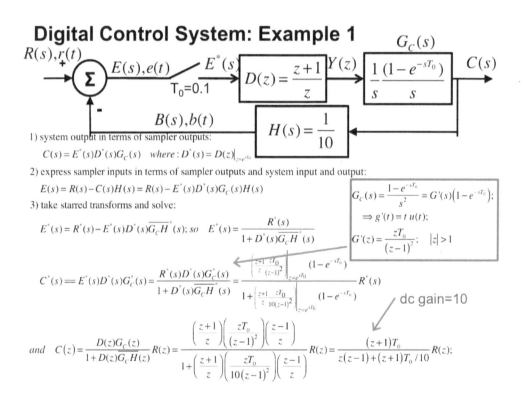

Digital Control System: Example 1

1) system output in terms of sampler outputs:

$C(s) = E^*(s)D^*(s)G_C(s)$ where: $D^*(s) = D(z)\big|_{z=e^{sT_0}}$

2) express sampler inputs in terms of sampler outputs and system input and output:

$E(s) = R(s) - C(s)H(s) = R(s) - E^*(s)D^*(s)G_C(s)H(s)$

3) take starred transforms and solve:

$E^*(s) = R^*(s) - E^*(s)D^*(s)\overline{G_C H}^*(s)$; so $E^*(s) = \dfrac{R^*(s)}{1 + D^*(s)\overline{G_C H}^*(s)}$

$G_C(s) = \dfrac{1 - e^{-sT_0}}{s^2} = G'(s)(1 - e^{-sT_0})$;

$\Rightarrow g'(t) = t\, u(t)$;

$G'(z) = \dfrac{zT_0}{(z-1)^2}$; $|z| > 1$

$C^*(s) = E^*(s)D^*(s)G^*_C(s) = \dfrac{R^*(s)D^*(s)G^*_C(s)}{1 + D^*(s)\overline{G_C H}^*(s)} = \dfrac{\left.\frac{z+1}{z}\frac{zT_0}{(z-1)^2}\right|_{z=e^{sT_0}}(1-e^{-sT_0})}{1 + \left.\frac{z+1}{z}\frac{zT_0}{10(z-1)^2}\right|_{z=e^{sT_0}}(1-e^{-sT_0})}R^*(s)$

dc gain=10

and $C(z) = \dfrac{D(z)G_C(z)}{1 + D(z)\overline{G_C H}(z)}R(z) = \dfrac{\left(\frac{z+1}{z}\right)\left(\frac{zT_0}{(z-1)^2}\right)\left(\frac{z-1}{z}\right)}{1 + \left(\frac{z+1}{z}\right)\left(\frac{zT_0}{10(z-1)^2}\right)\left(\frac{z-1}{z}\right)}R(z) = \dfrac{(z+1)T_0}{z(z-1) + (z+1)T_0/10}R(z)$;

Digital Control System: Example 1 using $G_{CL}(z)$

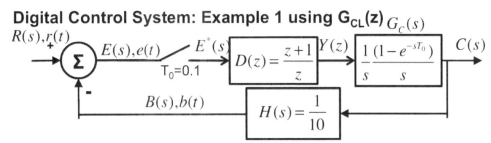

- For the system, check stability using characteristic equation:

$$G_{CL}(z)=\frac{C(z)}{R(z)}=\frac{D(z)G_c(z)}{1+D(z)G_cH(z)}=\frac{\left(\frac{z+1}{z}\right)\left(\frac{T_0}{(z-1)}\right)}{1+\left(\frac{z+1}{z}\right)\left(\frac{T_0}{10(z-1)}\right)}=\frac{(z+1)T_0}{z(z-1)+(z+1)T_0/10}$$

1) check stability from characteristic equation:

Characteristic equation

$$z(z-1)\left[1+\left(\frac{z+1}{z}\right)\left(\frac{T_0}{10(z-1)}\right)\right]=0 \Rightarrow 10z(z-1)+T_0(z+1)=0$$

$$\Rightarrow 10z^2-10z+T_0z+T_0=0 \Rightarrow z=\frac{10-T_0\pm\sqrt{(10-T_0)^2-40T_0}}{20}$$

$$\Rightarrow 0.01, 0.985$$

Stable roots of C(z) inside unit circle

Bilinear Transform Stability Analysis Method

Stability Analysis: Bilinear Transform Methods

- Bilinear transform methods
- Basic idea:
    - Map z-plane into some sort of s-plane (warped)
    - Use s-plane methods to determine stability
- Particular method used here:
    - Bilinear transform
    - Also known as w-transform or w-transformation
    - The w-plane is similar to s-plane, in that stable systems have poles in the left-half w-plane
    - However, frequency axis/mapping is somehow distorted or warped (through tangent function)

Bilinear Transform (w-transform)

- Consider the bilinear transform used to convert a continuous-time filter $H_c(s)$ into a discrete-time filter $H(z)$

$$H(z) = H_c(s)\Big|_{s=\frac{2}{T_0}\left(\frac{z-1}{z+1}\right)=\frac{2}{T_0}\left(\frac{1-z^{-1}}{1+z^{-1}}\right)} \quad where: \quad \sigma = \frac{2}{T_0}\frac{r^2-1}{1+r^2+2r\cos(\omega)} \qquad \Omega = \frac{2}{T_0}\frac{2r\sin(\omega)}{1+r^2+2r\cos(\omega)}$$

on unit circle: $\Omega = \frac{2}{T_0}\frac{\sin(\omega)}{1+\cos(\omega)} = 2\tan(\omega/2)/T_0 \quad or: \quad \omega = 2\tan^{-1}(\Omega T_s/2)$

- The bilinear transform maps left-half s-plane inside z-plane unit circle
- But frequencies are warped through tangent function
- Rearranging and solving for z as an "inverse bilinear transform" gives

$$s = \frac{2}{T_0}\left(\frac{z-1}{z+1}\right) \qquad \Rightarrow \qquad z = \left(\frac{2/T_0+s}{2/T_0-s}\right) = \left(\frac{1+sT_0/2}{1-sT_0/2}\right)$$

- Because frequencies are warped, rename the "s" variable to "w" to distinguish it from the unwarped "s" variable of the s-plane
- Thus, the w-transform maps the inside of the z-plane unit circle to the left-half w-plane

$$z = \left(\frac{2/T_0+w}{2/T_0-w}\right) = \left(\frac{1+wT_0/2}{1-wT_0/2}\right)$$

Text refers to this as bilinear transform

Stability Analysis Using W-Transform

$$H(w) = H_z(z)\Big|_{z=\left(\frac{1+wT_0/2}{1-wT_0/2}\right)} \quad where: \quad w = \sigma_w + j\Omega_w \quad and \quad s = \sigma + j\Omega$$

$$\sigma_w = \frac{2}{T_0}\frac{r^2-1}{1+r^2+2r\cos(\omega)} \qquad \Omega_w = \frac{2}{T_0}\frac{2r\sin(\omega)}{1+r^2+2r\cos(\omega)} \quad where: \quad z = re^{j\omega} = e^{sT_0} = e^{(\sigma+j\Omega)T_0}$$

on unit circle: $\Omega_w = 2\tan(\omega/2)/T_0 = 2\tan(\Omega T_0/2)/T_0 \quad or: \Omega T_0 = 2\tan^{-1}(\Omega_w T_s/2)$

- The w-transform:
  - Definition: $H(w)=H_z(z)|_{s=(1+wTo/2)/(1-wTo/2)}$
  - Maps the inside of the unit circle in the z-plane to left-half w-plane
  - Maps the unit circle in the z-plane to the w-plane imaginary axis
  - But frequencies are warped through tangent function
- Key item: stable poles in z map into stable poles in w (treating w like s)
  - Noting that w is not the same as s due to frequency warping
  - Despite warping, stable poles are located in left half w-plane
- After w-transformation:
  - Can use continuous-time tools to analyze stability
    - Routh-Hurwitz
    - Jury

Example 1: W-Transform

$$H_z(z) = \frac{(z+1)^2}{(z+0.5)(z-0.5)} \quad and \quad T_0 = 1$$

$$H(w) = H_z(z)\Big|_{z=\left(\frac{1+wT_0/2}{1-wT_0/2}\right)} = \frac{\left(\frac{1+wT_0/2}{1-wT_0/2}+1\right)^2}{\left(\frac{1+wT_0/2}{1-wT_0/2}+0.5\right)\left(\frac{1+wT_0/2}{1-wT_0/2}-0.5\right)}$$

$$= \frac{64}{(2+3wT_0)(6+wT_0)} = \frac{64}{3(2/3+w)(6+w)}$$

- The w-transform:
  - Definition: $H(w)=H_z(z)|_{s=(1+wTo/2)/(1-wTo/2)}$
- Key item: stable poles in z map into stable poles in w (treating w like s)
  - In example:
    - Stable z-plane poles at z= 0.5 and -0.5
      $\Rightarrow$map to stable w-plane poles at w= -0.67 and -6
    - Also z-plane double zero at z= -1
      $\Rightarrow$Map to 2 zeroes at w = ∞
  - Despite warping, stable z-pole maps to stable w-pole

Example 2: W-Transform

$$H_z(z) = \frac{(z-1)^2}{(z^2+1)} = \frac{(z-1)^2}{(z+j)(z-j)} \quad and \quad T_0 = 1/3$$

$$H(w) = H_z(z)\Big|_{z=\left(\frac{1+wT_0/2}{1-wT_0/2}\right)} = \frac{\left(\frac{1+wT_0/2}{1-wT_0/2}-1\right)^2}{\left(\frac{1+wT_0/2}{1-wT_0/2}\right)^2+1}$$

$$= \frac{2w^2}{(36+w^2)} = \frac{2w^2}{(w+j6)(w-j6)}$$

- The w-transform:
  - Map $H(w)=H_z(z)|_{s=(1+wTo/2)/(1-wTo/2)}$
- Key item: stable poles in z map into stable poles in w (treating w like s)
  - In example:
    - Unstable z-plane poles at z= j and -j
      $\Rightarrow$Maps to unstable w-plane poles at w= j6 and -j6
    - Also z-plane zeroes at z= 1
      $\Rightarrow$Maps to zeroes at w = 0
  - Despite warping, maps z-circle to jw-axis

Routh-Hurwitz Criteria

- Routh-Hurwitz:
    - ○ Method to determine if roots of polynomial are in left-half plane
    - ○ Works for s-plane or w-plane
- Method:
    - ○ Form Routh table (Routhian array)
        - – First row: highest power coefficient, then every other
        - – Second row: next-highest coefficient, then every other
        - – Subsequent rows: $(b_0 a_R - a_0 b_R)/b_0$
            - a_0 is upper row leftmost, a_R is upper row right
            - b_0 is lower row leftmost, b_R is lower row right
    - ○ Number of sign changes in left column = number of unstable poles in open right-half plane

Routh-Hurwitz: Example 1

- Example: $H(w) = w^4 - 2w^3 - 13 w^2 + 38 w - 24$
- Form Routh table (Routhian array): $(b_0 a_R - a_0 b_R)/b_0$

$$
\begin{array}{ccc}
1 & -13 & -24 \\
-2 & 38 & 0 \\
6 & -24 & 0 \\
30 & 0 & 0 \\
-24 & 0 & 0 \\
\end{array}
$$

- So, the system is unstable
    - ○ with 3 sign changes indicating 3 unstable poles
- For this example: $H(w)=(w-1)(w-2)(w-3)(w+4)$

Routh-Hurwitz: Example 2

- Example: $H(w) = w^3 + 6 w^2 + 11 w + 6$
- Form Routh table (Routhian array)

1	11
6	6
10	0
6	0

- So, the system is stable
    - With NO sign changes indicating NO unstable poles
- For this example: $H(w)=(w+1)(w+2)(w+3)$

Stability Analysis: Jury Test

- Jury test may be applied directly to H(z)
- Very complicated
- Said to be most useful for H(z) of less than 3rd order
    - But, can easily use quadratic formula for 2nd order
- Displaced by modern computer methods

Stability Analysis: Modern Methods

- Computer methods :
  - Symbolic math & solvers to find roots (poles & zeroes)
  - Numerical solvers to find roots (poles & zeroes)
- System poles and zeroes determined on computer

- So:
  - Less need for Jury method & Routh-Hurwitz method
  - ...but good for exams!

- Good example of computer methods:
  - Root locus methods

Root Locus Stability Analysis Methods

Root Locus Methods

- Root locus stability analysis methods
    - System poles and zeroes determined on computer
        - Manual methods too: asymptotes, breakaway points
        - We will focus on computer methods
        - But beware pole-zero cancellations
    - Typically focused on system poles
        - Poles are zeroes of characteristic equation
    - Could include zeroes for more information

- Basic idea:
    - Plot system poles as a function of some parameter
    - Parameter is typically a gain factor K in the loop
        - Parameter does not have to be gain!
        - Could be R, L, C, etc.

Root Locus: using configuration 3

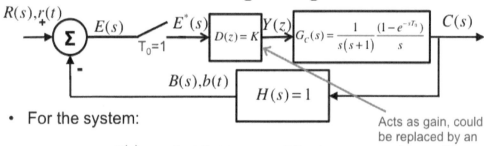

- For the system:

$$G_{CL}(z) = \frac{C(z)}{R(z)} = \frac{D(z)G_C(z)}{1 + D(z)G_C H(z)} = \frac{KG_C(z)}{1 + KG_C H(z)}$$

Acts as gain, could be replaced by an analog amplifier

- Where D(z)=K acts as an amplifier with gain K
- Root Locus is a plot of poles of $G_{CL}(z)$ as function of gain K
- Poles of $G_{CL}(z)$ are zeroes of characteristic equation

Root Locus Example 1

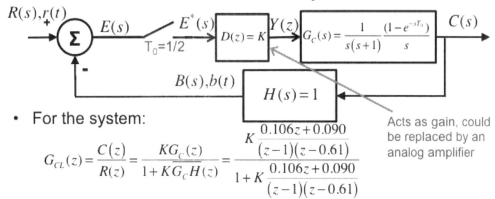

- For the system:

$$G_{CL}(z) = \frac{C(z)}{R(z)} = \frac{KG_c(z)}{1+KG_cH(z)} = \frac{K\dfrac{0.106z+0.090}{(z-1)(z-0.61)}}{1+K\dfrac{0.106z+0.090}{(z-1)(z-0.61)}}$$

Acts as gain, could be replaced by an analog amplifier

- Characteristic equation is

$$0 = (z-1)(z-0.61)\left(1+K\frac{0.106z+0.090}{(z-1)(z-0.61)}\right)$$

$$or: \quad (z-1)(z-0.61)+K(0.106z+0.090)=0$$

$$z^2+(0.106K-1.61)z+0.61+0.09K=0$$

- Root locus is a plot of zeroes of characteristic equation

(Adapted from Phillips & Nagle, ISBN 978-0132938310, Ex. 7.7)

Root Locus Example 1

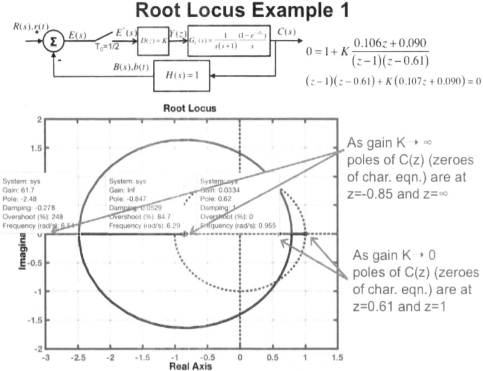

Root Locus Example 2

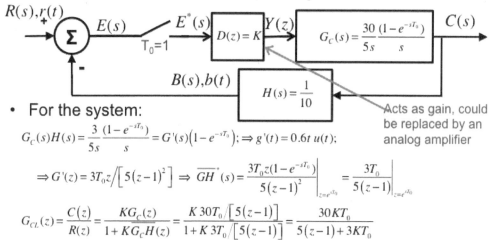

- For the system:

$$G_c(s)H(s) = \frac{3}{5s}\frac{(1-e^{-sT_0})}{s} = G'(s)(1-e^{-sT_0}); \Rightarrow g'(t) = 0.6t\,u(t);$$

$$\Rightarrow G'(z) = 3T_0 z \Big/ \Big[5(z-1)^2\Big] \Rightarrow \overline{GH}^*(s) = \frac{3T_0 z(1-e^{-sT_0})}{5(z-1)^2}\bigg|_{z=e^{sT_0}} = \frac{3T_0}{5(z-1)}\bigg|_{z=e^{sT_0}}$$

$$G_{CL}(z) = \frac{C(z)}{R(z)} = \frac{KG_c(z)}{1+KG_cH(z)} = \frac{K\,30T_0\Big/\Big[5(z-1)\Big]}{1+K\,3T_0\Big/\Big[5(z-1)\Big]} = \frac{30KT_0}{5(z-1)+3KT_0}$$

- Characteristic equation is

$$5(z-1)+3KT_0 = 0 = 5z+3KT_0-5 = 5z+3K-5$$

Root Locus Example 2:

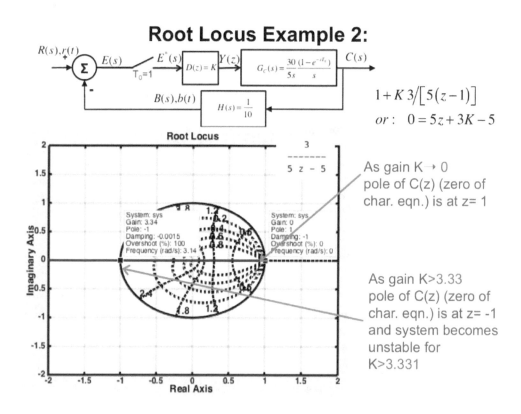

$$1+K\,3\Big/\Big[5(z-1)\Big]$$

$$or: \quad 0 = 5z+3K-5$$

As gain $K \to 0$
pole of C(z) (zero of
char. eqn.) is at z= 1

As gain K>3.33
pole of C(z) (zero of
char. eqn.) is at z= -1
and system becomes
unstable for
K>3.331

Admittance Matrix Analysis

- Many systems are modeled by a circuit or contain circuits
- Admittance matrices are a simple way to analyze circuits
- To construct the admittance matrix Y:
  - Construct a column vector of the voltages at every node
  - Along diagonal of Y, y_{nn}=sum of admittances touching v_{nn}
  - Rest of row in Y, y_{mn}= negative of admittances touching v_m and v_n
  - Vector I =current sources to ground connected to circuit

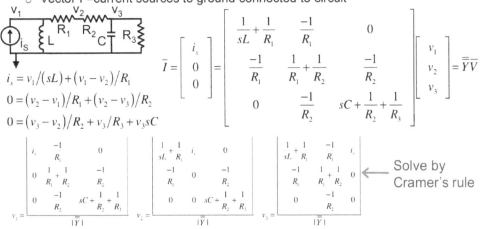

$$\bar{I} = \begin{bmatrix} i_s \\ 0 \\ 0 \end{bmatrix} = \begin{bmatrix} \dfrac{1}{sL}+\dfrac{1}{R_1} & \dfrac{-1}{R_1} & 0 \\[2mm] \dfrac{-1}{R_1} & \dfrac{1}{R_1}+\dfrac{1}{R_2} & \dfrac{-1}{R_2} \\[2mm] 0 & \dfrac{-1}{R_2} & sC+\dfrac{1}{R_2}+\dfrac{1}{R_3} \end{bmatrix} \begin{bmatrix} v_1 \\ v_2 \\ v_3 \end{bmatrix} = \bar{\bar{Y}}\,\bar{V}$$

$i_s = v_1/(sL)+(v_1-v_2)/R_1$

$0 = (v_2-v_1)/R_1+(v_2-v_3)/R_2$

$0 = (v_3-v_2)/R_2+v_3/R_3+v_3sC$

$$v_1 = \frac{\begin{vmatrix} i_s & \dfrac{-1}{R_1} & 0 \\[2mm] 0 & \dfrac{1}{R_1}+\dfrac{1}{R_2} & \dfrac{-1}{R_2} \\[2mm] 0 & \dfrac{-1}{R_2} & sC+\dfrac{1}{R_2}+\dfrac{1}{R_3} \end{vmatrix}}{|\bar{\bar{Y}}|}$$

$$v_2 = \frac{\begin{vmatrix} \dfrac{1}{sL}+\dfrac{1}{R_1} & i_s & 0 \\[2mm] \dfrac{-1}{R_1} & 0 & \dfrac{-1}{R_2} \\[2mm] 0 & 0 & sC+\dfrac{1}{R_2}+\dfrac{1}{R_3} \end{vmatrix}}{|\bar{\bar{Y}}|}$$

$$v_3 = \frac{\begin{vmatrix} \dfrac{1}{sL}+\dfrac{1}{R_1} & \dfrac{-1}{R_1} & i_s \\[2mm] \dfrac{-1}{R_1} & \dfrac{1}{R_1}+\dfrac{1}{R_2} & 0 \\[2mm] 0 & \dfrac{-1}{R_2} & 0 \end{vmatrix}}{|\bar{\bar{Y}}|}$$

← Solve by Cramer's rule

9 DIGITAL LAG CONTROLLER DESIGN

This chapter provides discussion of the design of digital lag compensators for digital control systemss.

Digital Lag Controller Design

Digital Lag Controller Design

- Last time:
  - o Stability of closed-loop discrete-time systems
    - – Poles and zeroes of $G_{CL}(z)$
    - – Zeroes of characteristic equation
    - – And beware of pole/zero cancellation
  - o Analysis methods
    - – Root locus, Routh-Hurwitz, others
- Next:
  - o Bode analysis: phase and gain margin
    - – In context of digital controller design, D(z)
    - – Find D(z) to assure stability by setting gain and phase margin
  - o Controller design types
    - – Digital phase-lag compensation
    - – Digital phase-lead compensation
    - – Digital lag-lead compensation
    - – Digital PID (proportional-integral-derivative) compensation

Discrete-Time Control System Design Goals

- Discrete-time control system design
    - Main goal: design stable D(z) to meet some overall system design goal
- Types of overall system design goals (may be one or more)
    - Stability margin (closed loop): gain and phase margin
    - Steady state error/accuracy (closed loop):
        - Step u[n] (position) error: $\text{Lim}_{n \to \infty} e[n] = \text{Lim}_{z \to 1} (z-1)E(z)$
        - Ramp n u[n] (velocity) error: $\text{Lim}_{n \to \infty} e[n] = \text{Lim}_{z \to 1} (z-1)E(z)/T_0$
    - Transient response (closed loop):
        - Rise time, overshoot, damping, settling time (dominant pole/poles)
    - Bandwidth (closed loop): inversely proportional to rise time
    - Sensitivity to component parameter variations for devices in loop
    - Rejection of disturbance/noise signal added at any point in loop
    - Control effort: maximum limits on torque, energy, voltage clipping
- Next:
    - Focus on closed-loop stability margin: gain and phase margin

Discrete-Time Control System Design Goals

$$G_C(s) = \frac{1}{s(s+1)} \qquad H(s) = 1 \qquad T_0 = 1 \qquad G_{CL}(s) = \frac{1}{s(s+1)+1}$$

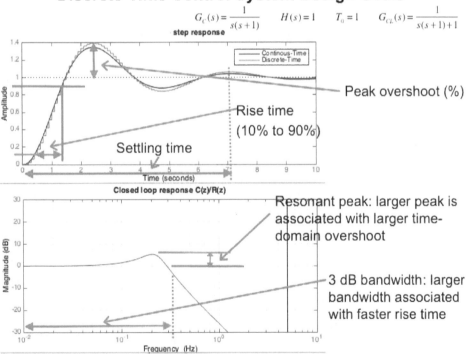

Peak overshoot (%)

Rise time (10% to 90%)

Settling time

Resonant peak: larger peak is associated with larger time-domain overshoot

3 dB bandwidth: larger bandwidth associated with faster rise time

Discrete-Time Control System Design Goals

$$G_c(s) = \frac{1}{s(s+1)} \qquad H(s) = 1 \qquad T_0 = 1 \qquad G_{CL}(s) = \frac{1}{s(s+1)+1}$$

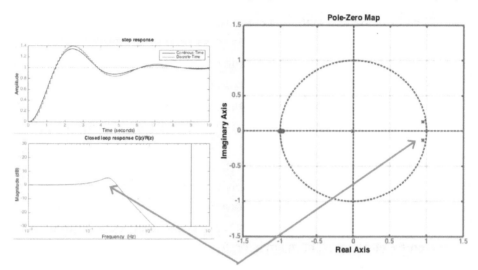

3 dB bandwidth and resonant peak
associated with pole locations

Review of Continuous-Time Bode Plots

Example: Amplifier Feedback

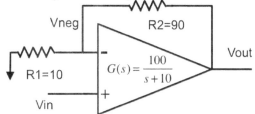

- Recall: no current into op-amp terminals, so
  - Vneg/R1 = Vout/(R1+R2)
  - Vout(s) = (Vin-Vneg(s)) G(s)
  - Resistors set feedback H(s)

$$V_{out}/(R_1+R_2) = V_{neg}/R_1 = (V_{in}-V_{out}/G(s))/R_1$$

$$\frac{V_{out}(s)}{V_{in}(s)} = \frac{G(s)}{1+G(s)R_1/(R_1+R_2)} = \frac{G(s)}{1+G(s)H(s)}$$

$$G_{CL}(s) = \frac{V_{out}(s)}{V_{in}(s)} = \frac{100/(s+10)}{1+1000/(s+10)/100} = \frac{100}{s+20} = 5 @ s = 0$$

Example: Amplifier Feedback Bode Magnitude Plot

$$G_{cL}(s) = \frac{V_{out}(s)}{V_{in}(s)} = \frac{G(s)}{1+G(s)H(s)} = \frac{100}{s+20}$$

$$G(s) = \frac{100}{s+10} \quad G(s)H(s) = \frac{10}{s+10} \quad G_c(s) = \frac{100}{s+20}$$

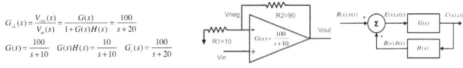

- Plot $|G(s)|$, $|G(s)H(s)|$, and closed-loop gain $|G_{CL}(s)|$ in dB
  - Recall: Bode magnitude breaks 20dB/decade at 3dB point

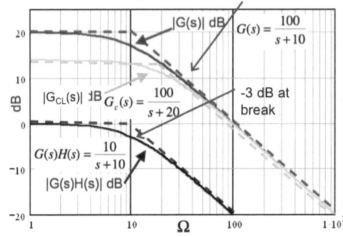

$|G(s)| \text{ dB}$ $G(s) = \dfrac{100}{s+10}$

$|G_{CL}(s)| \text{ dB}$ $G_c(s) = \dfrac{100}{s+20}$

-3 dB at break

$G(s)H(s) = \dfrac{10}{s+10}$

$|G(s)H(s)| \text{ dB}$

Bode plot analysis of previous example

Example: Amplifier Feedback Bode Phase Plot

$$G_{cL}(s) = \frac{V_{out}(s)}{V_{in}(s)} = \frac{G(s)}{1+G(s)H(s)} = \frac{100}{s+20}$$

$$G(s) = \frac{100}{s+10} \quad G(s)H(s) = \frac{10}{s+10} \quad G_c(s) = \frac{100}{s+20}$$

- Plot $\angle G(s)$, $\angle G(s)H(s)$, and closed-loop phase $\angle G_{CL}(s)$
  - Recall: Phase (degrees) is -45° at 3dB point, -45°/decade

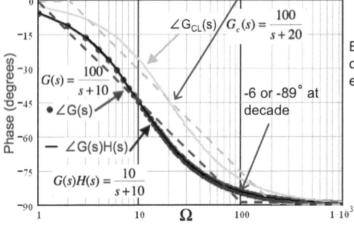

$\angle G_{CL}(s)$ $G_c(s) = \dfrac{100}{s+20}$

$G(s) = \dfrac{100}{s+10}$

$\bullet \angle G(s)$

-6 or -89° at decade

$- \angle G(s)H(s)$

$G(s)H(s) = \dfrac{10}{s+10}$

Bode plot analysis of previous example

Bode Phase/Gain Margin Design Methods

Bode: Minimum-Phase Caution

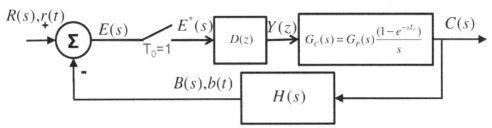

- Bode analysis considers open-loop transfer function $G_{OL}(z)$:

$$G_{OL}(z) = D(z)\left(\overline{G_C H}^*(s)\right)\Big|_{e^{sT}=z} = D(z)\overline{G_C H}(z) \qquad \text{for the example system above}$$

- However, caution must be taken if $G_{OL}(z)$ is not minimum phase
  - It is better to use Nyquist plot for such cases
- Also, beware any pole-zero cancelations in $G_{OL}(z)$
- A minimum-phase polynomial $G_{OL}(z)$:
  - Has all poles and zeroes inside unit circle
  - Has no poles on unit circle (some authors "allow" a single pole at z=1)

Bode Gain & Phase Margin (Configuration 3)

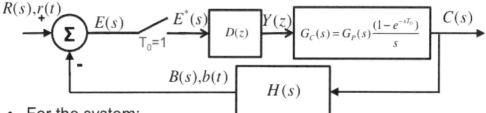

- For the system:

$$G_{CL}(z) = \frac{C(z)}{R(z)} = \frac{D(z)G_C(z)}{1 + D(z)\overline{G_C H}(z)} = \frac{D(z)G_C(z)}{1 + G_{OL}(z)}$$

- Goal design D(z) to meet overall system design goals of:
  - Gain margin: dB $|G_{OL}(z)|$ is less than 0 dB at $\angle G_{OL}(z) = -180°$
  - Phase margin: degrees $\angle G_{OL}(z) > -180°$ at $|G_{OL}(z)| = 1$
- Reasoning for this approach:
  - Because poles of $G_{CL}(z)$ are are at $G_{OL}(z) = -1 = 1\angle -180°$

  due to the denominator being:

$$1 + K\overline{G_C H}(z) = 1 + G_{OL}(z) = 0$$

Example 1: Continuous-Time Bode Gain & Phase Margin

$$G_c(s) = \frac{10^7}{(s+10)(s+1000)} \qquad H(s) = \frac{1}{10} \qquad G_{CL}(s) = \frac{10^7}{(s+10)(s+1000)+10^6}$$

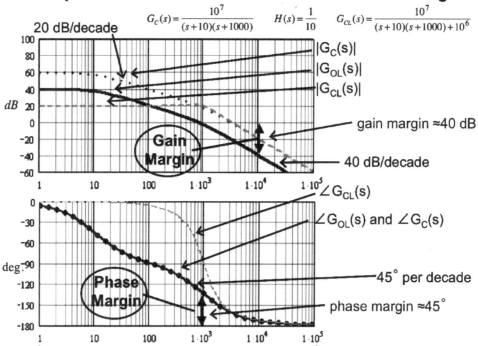

Example 2: Continuous-Time Bode Gain & Phase Margin

$$G_C(s) = \frac{10^8}{(s+10)(s+1000)} \qquad H(s) = \frac{1}{10} \qquad G_{CL}(s) = \frac{10^8}{(s+10)(s+1000)+10^7}$$

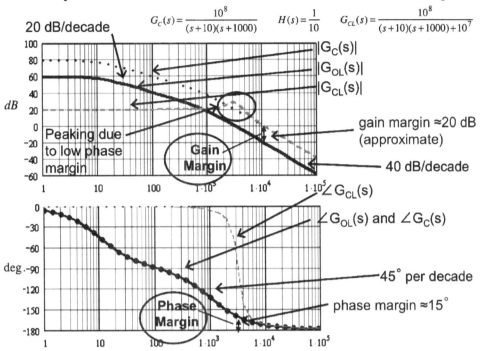

Bode Phase/Gain Margin Design Methods

Bode Methods: Digital Phase-Lag Compensation

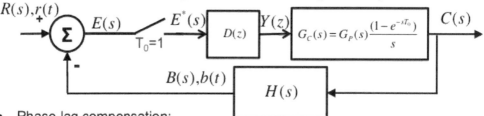

- Phase-lag compensation:
  - Reduces high frequency gain to improve gain phase margin
  - Introduces some phase lag at low/intermediate frequency
- General form for phase-Lag in z-domain and w-domain :

$$D(z) = K_D \frac{z - z_0}{z - z_P} \implies D(w) = D(z)\Big|_{z = \frac{1 + wT_0/2}{1 - wT_0/2}} = K_D \frac{w - (2/T_0)\left[(z_0 - 1)/(z_0 + 1)\right]}{w - (2/T_0)\left[(z_P - 1)/(z_P + 1)\right]}$$

$$D(z) = a_0 \frac{1 + w/w_0}{1 + w/w_P}\Big|_{w = \frac{2}{T_0}\frac{z-1}{z+1}} = \left(a_0 \frac{w_P(1 + w_0 T_0/2)}{w_0(1 + w_P T_0/2)}\right) \frac{z - (1 - w_0 T_0/2)/(1 + w_0 T_0/2)}{z - (1 - w_P T_0/2)/(1 + w_P T_0/2)}$$

where : $w_0 > w_P$ for a phase-lag compensator, and ← ———— NOTE

$w = \sigma_w + j\Omega_w$ and $z = e^{sT_0}$ and $\Omega_w T_0 = 2\tan(\omega/2) = 2\tan(\Omega T_0/2)$

- Design of compensator D(z) is done in w-domain

Phase Lag Frequency Response in w-domain

- Phase-lag compensator response characteristics:
  - a_0 sets dc gain, $w_0 > w_P$
  - max phase lag at midpoint of w_0 and w_P
  - Phase approx. 5 degrees at 10 w_0

$$D(w) = a_0 \frac{1 + w/w_0}{1 + w/w_P}$$

$$where: w = \sigma_w + j\Omega_w;$$

$$\Omega_w T_0 = 2\tan(\Omega T_0 / 2)$$

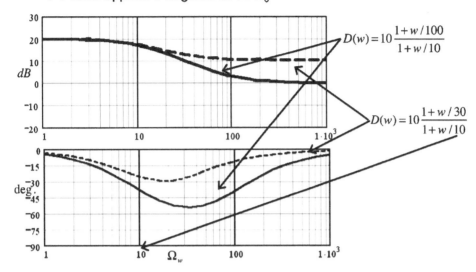

$$D(w) = 10\frac{1 + w/100}{1 + w/10}$$

$$D(w) = 10\frac{1 + w/30}{1 + w/10}$$

Phase-Lag Bode Plots in w-domain

- Phase-lag compensator response characteristics:
  - a_0 sets dc gain, $w_0 > w_P$
  - max phase lag at midpoint of w_0 and w_P
  - Phase approx. 5 degrees at 10 w_0

$$D(w) = a_0 \frac{1 + w/w_0}{1 + w/w_P}$$

$$where: w = \sigma_w + j\Omega_w;$$

$$\Omega_w T_0 = 2\tan(\Omega T_0 / 2)$$

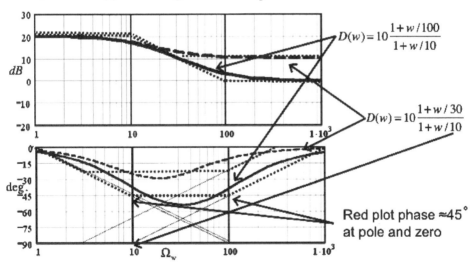

$$D(w) = 10\frac{1 + w/100}{1 + w/10}$$

$$D(w) = 10\frac{1 + w/30}{1 + w/10}$$

Red plot phase $\approx 45°$
at pole and zero

Example Phase-Lag Bode Plots in w-domain

- Phase-lag compensator Bode plot analysis example $\qquad D(w) = 1\dfrac{1+w/100}{1+w/10}$

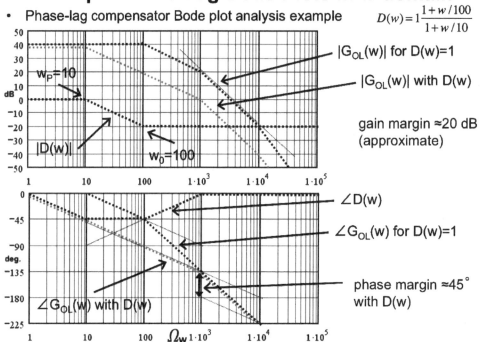

$|G_{OL}(w)|$ for D(w)=1

$|G_{OL}(w)|$ with D(w)

gain margin ≈20 dB (approximate)

w_P=10

|D(w)|

w_0=100

∠D(w)

∠$G_{OL}(w)$ for D(w)=1

phase margin ≈45° with D(w)

∠$G_{OL}(w)$ with D(w)

Digital Phase-Lag Compensation Procedure

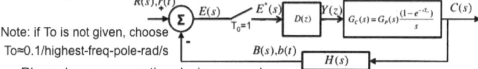

Note: if To is not given, choose
To≈0.1/highest-freq-pole-rad/s

- Phase-lag compensation design procedure:
  1. Take w-transform of open-loop gain $G_{OL}(z)$ with D(z)=1 to get $G_{OL}(w)$
  2. Decide upon a phase margin ϕ (default ϕ=45 degrees)
  3. Define frequency $w_1 = j\Omega_{W1}$ where $\angle G_{OL}(w_1) = -180+\phi+5$ degrees
  4. Set the zero of D(w) using $w_0 = \Omega_{W0} = \text{Im}\{w_1\}/10$ where $\Omega_{W0} = \Omega_{W1}/10$
  5. Set a_0 such that the desired dc gain G_{OLdc} is obtained
     - Gain determined by other specs such as steady-state error, etc
     $$G_{OLdc} = D(w)\overline{G_C H}(w)\Big|_{w=0} = a_0\overline{G_C H}(w)\Big|_{w=0} \quad \Rightarrow \quad a_0 = G_{OLdc}\big/\overline{G_C H}(w)\Big|_{w=0}$$
  6. Set the open-loop unity gain point at w_1: $D(w_1) G_{OL}(w_1)=1 \implies w_P$
     $$\left|D(w_1)G_{OL}(w_1)\right| = \left|a_0\frac{1+w_1/w_0}{1+w_1/w_P}G_{OL}(w_1)\right| \approx \left|a_0\frac{w_P}{w_0}G_{OL}(w_1)\right| = 1 \quad so \quad w_P \approx \left|\frac{w_0}{a_0 G_{OL}(w_1)}\right|$$
  7. Convert back to z-domain using w_0, w_P, and a_0 from above

  Note: w_0 and w_P are real-valued

  $$D(z) = K_D\frac{z-z_0}{z-z_p} = \left(a_0\frac{w_P(1+w_0 T_0/2)}{w_0(1+w_P T_0/2)}\right)\frac{z-(1-w_0 T_0/2)\big/(1+w_0 T_0/2)}{z-(1-w_P T_0/2)\big/(1+w_P T_0/2)}$$

Example 1: Digital Phase-Lag Compensation.

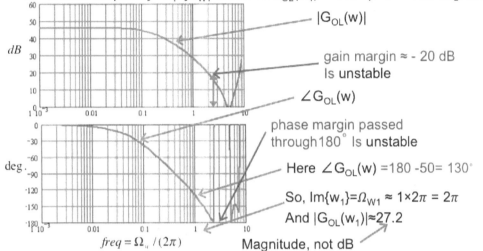

- Phase-lag design procedure:
    1. Take w-transform of open-loop gain $G_{OL}(z)$ with $D(z)=1$ to get $G_{OL}(w)$
    2. Decide upon a phase margin ϕ (default $\phi=45$ degrees)
    3. Define frequency $w_1=j\Omega_{W1}$ where $\angle G_{OL}(w_1)= -180+\phi+5 \approx -130$ degrees

— $|G_{OL}(w)|$

gain margin $\approx$ - 20 dB
Is unstable

— $\angle G_{OL}(w)$

phase margin passed
through180° Is unstable

Here $\angle G_{OL}(w) =180 -50= 130°$

So, $Im\{w_1\}=\Omega_{W1} \approx 1\times2\pi = 2\pi$
And $|G_{OL}(w_1)|\approx27.2$

Magnitude, not dB

Example 1: Digital Phase-Lag Compensation

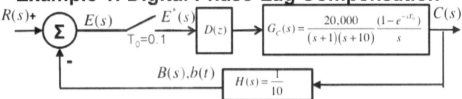

- Phase-lag design procedure, cont'd:
    3. From previous slide, $w_1= j\, 2\,\pi$
    4. Set the zero of D(w) using $w_0=\Omega_{W0}= Im\{w_1\}/10$ where $\Omega_{W0}=\Omega_{W1}/10$
        So: $w_0 = Re\{w_1\}/10 = 0.2\,\pi$
    5. Set a_0 such that the desired dc gain $G_{OL}(w)|_{w=0}$ is obtained
        - Gain determined by other specs such as steady-state error, gain, etc
        - Here choose dc gain $D(z)_{z=1} = a_0 = 0.1$ and look at steady-state error

$$G_{CL}(z)=\frac{D(z)G_C(z)}{1+D(z)\overline{G_C H}(z)} \quad and \quad \frac{E(z)}{R(z)}=\frac{1}{1+D(z)\overline{G_C H}(z)} \quad choose: D(1)=0.1=a_0$$

$$so: G_{CL}(1)=\frac{D(1)G_C(1)}{1+D(1)\overline{G_C H}(1)}=\frac{200}{21}=9.52 \quad and \quad \frac{E(1)}{R(1)}=\frac{1}{1+D(1)\overline{G_C H}(1)}=\frac{1}{21}=0.048$$

Example 1: Digital Phase-Lag Compensation

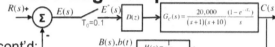

- Phase-lag design procedure, cont'd:

  6. Set the open-loop unity gain point at w_1: $D(w_1)\,G_{OL}(w_1)=1$ yields w_P

$$w_P \approx \left|\frac{w_0}{a_0 G_{OL}(w_1)}\right| = \left|\frac{0.2\pi}{0.1\times 27.2}\right| = 0.23$$

  7. Convert back to z-domain using w_z, w_P, and a_0 from above

$$D(z)=K_D\frac{z-z_0}{z-z_P}=\left(a_0\frac{w_P\left(1+w_0T_0/2\right)}{w_0\left(1+w_PT_0/2\right)}\right)\frac{z-\left(1-w_0T_0/2\right)/\left(1+w_0T_0/2\right)}{z-\left(1-w_PT_0/2\right)/\left(1+w_PT_0/2\right)} = 0.037\frac{z-0.939}{z-0.977}$$

  8. And is a good idea to recheck closed loop zeroes and poles

      2 zeroes at z=0.9390, and -0.6945

      3 poles at

      z=0.9407 + 0.0000i, 0.5232 + 0.5041i, and 0.5232 - 0.5041i

Example 1: Digital Phase-Lag Compensation

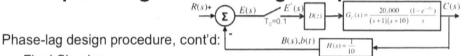

- Phase-lag design procedure, cont'd:
  o Final Checks:
    – Check phase and gain margin in new compensated $G_{CL}(z)$

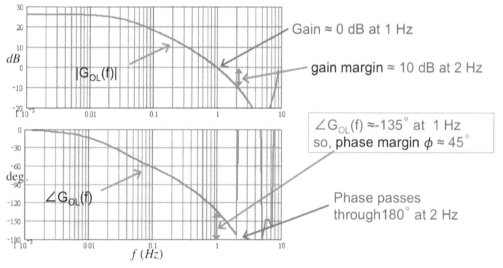

Gain ≈ 0 dB at 1 Hz

gain margin ≈ 10 dB at 2 Hz

$\angle G_{OL}(f) \approx -135°$ at 1 Hz
so, phase margin $\phi \approx 45°$

Phase passes through 180° at 2 Hz

Example 1: Digital Phase-Lag Compensation

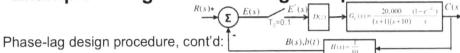

- Phase-lag design procedure, cont'd:
  - Final Checks:
    - Check phase and gain margin in new compensated $G_{CL}(z)$

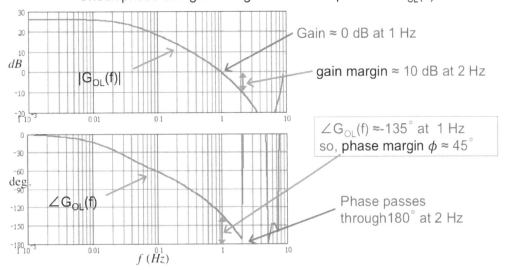

Gain ≈ 0 dB at 1 Hz

gain margin ≈ 10 dB at 2 Hz

$\angle G_{OL}(f) \approx -135°$ at 1 Hz
so, **phase margin** $\phi \approx 45°$

Phase passes through $180°$ at 2 Hz

$|G_{OL}(f)|$

$\angle G_{OL}(f)$

$f\ (Hz)$

Example 1: Digital Phase-Lag Compensation

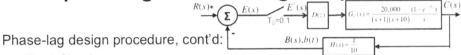

- Phase-lag design procedure, cont'd:
  - Final Checks:
    - Check phase and gain margin in new compensated $G_{CL}(z)$

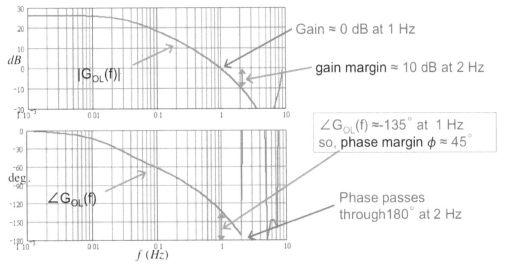

Gain ≈ 0 dB at 1 Hz

gain margin ≈ 10 dB at 2 Hz

$\angle G_{OL}(f) \approx -135°$ at 1 Hz
so, **phase margin** $\phi \approx 45°$

Phase passes through $180°$ at 2 Hz

$|G_{OL}(f)|$

$\angle G_{OL}(f)$

$f\ (Hz)$

Example 1: Digital Phase-Lag Compensation

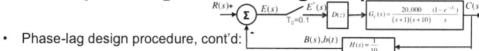

- Phase-lag design procedure, cont'd:
  - o Final Checks:
    - – Check phase and gain margin

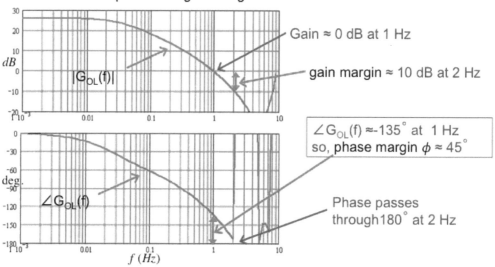

Gain ≈ 0 dB at 1 Hz

gain margin ≈ 10 dB at 2 Hz

$\angle G_{OL}(f) \approx -135°$ at 1 Hz
so, **phase margin** $\phi \approx 45°$

Phase passes
through 180° at 2 Hz

Summary: Digital Phase-Lag Compensation

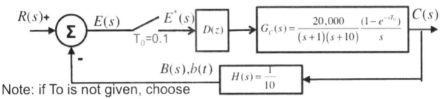

Note: if To is not given, choose

To≈0.1/highest-freq-pole in rad/s

- Phase-lag compensation design procedure:
    1. Take w-transform of open-loop gain $G_{OL}(z)$ with D(z)=1 to get $G_{OL}(w)$
    2. Decide upon a phase margin ϕ (default ϕ=45 degrees)
    3. Define frequency $w_1=j\Omega_{W1}$ where $\angle G_{OL}(w_1)$ = -180+ϕ+5 degrees
    4. Set the zero of D(w) using $w_0=\Omega_{W0}$= Im{w_1}/10   Note: w_0
    5. Set a_0 such that the desired dc gain G_{OLdc} is obtained   and w_p are
       real-valued
        - Gain determined by other specs such as steady-state error, etc
    6. Set the open-loop unity gain point at w_1: |D(w_1) $G_{OL}(w_1)$|=1 $\Rightarrow$ w_P
    7. Convert back to z-domain D(z) using w_Z, w_P, and a_0 from above
- Different control system configurations will have different details
- Beware numerical precision pole/zero shifts

Caution on Numerical Precision

- Numerical precision issues:
    o Pole and zero locations of D(z) may shift due to numerical precision
    o May induce instability
    o May cause a dc zero (no loop gain at dc!):
- Especially a problem in phase-lag compensator
    o Since w_P < w_0 < π/10, then poles and zeroes are often near z=1
    o Precision can cause unstable pole
    o Also, precision can cause zero to move to z=1 (zero dc gain!)
    o Also, smaller T_0 can exacerbate the problem, causing the pole and zero of D(z) to come closer to z=1

10 DIGITAL LEAD AND PID CONTROLLERS

This chapter covers the design and analysis of digital lead compensators and PID compensators for digital control systems.

Digital Lead and PID Controllers

Closed-Loop System Stability

- Last time:
  - o Digital phase-lag compensator
    - – Bode plot analysis
    - – Phase-lag compensator D(z) design
    - – Focus on closed-loop stability margin: gain and phase margin
  - o Other important closed-loop characteristics: steady state error for step and ramp, transient response rise time, overshoot, damping, settling time, bandwidth, sensitivity to component variation, rejection of disturbance/noise, control effort/maximum limits

- Next:
  - o Digital phase-lead compensator design
    - – Bode plot analysis
    - – Phase-lead compensator D(z) design
  - o Digital lag-lead compensator
  - o Digital PID controller (compensator) design
    - – PID compensator D(z) design

Bode Phase/Gain Margin Design Methods

Phase-Lead Frequency Response in w-domain

- Phase-lead compensator response characteristics:
  $$D(w) = a_0 \frac{1 + w/w_0}{1 + w/w_P}$$
  - a_0 sets dc gain, $w_P > w_0$
  - max phase lead at midpoint of w_0 and w_P
  - Phase approx. 5 degrees at 10 w_0

  $$where: w = \sigma_w + j\Omega_w;$$
  $$\Omega_w T_0 = 2\tan(\Omega T_0 / 2)$$

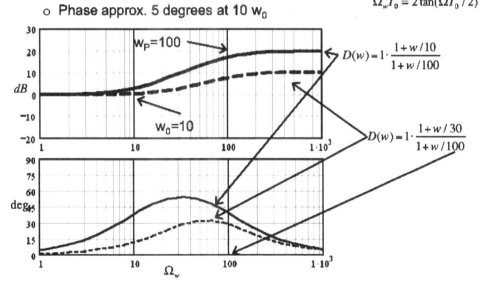

$$D(w) = 1 \cdot \frac{1 + w/10}{1 + w/100}$$

$$D(w) = 1 \cdot \frac{1 + w/30}{1 + w/100}$$

Phase-Lead Bode Plots in w-domain

- Phase-lead compensator response characteristics:

$$D(w) = a_0 \frac{1 + w/w_0}{1 + w/w_P}$$

$$where: w = \sigma_w + j\Omega_w;$$

$$\Omega_w T_0 = 2\tan(\Omega T_0 / 2)$$

  - a_0 sets dc gain, $w_P > w_0$
  - max phase lead at midpoint of w_0 and w_P
  - Phase approx. 5 degrees at 10 w_0

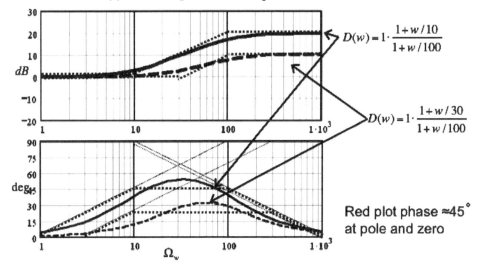

$$D(w) = 1 \cdot \frac{1 + w/10}{1 + w/100}$$

$$D(w) = 1 \cdot \frac{1 + w/30}{1 + w/100}$$

Red plot phase ≈45°
at pole and zero

Digital Phase-Lead Compensation Procedure

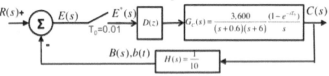

Note: if To is not given, choose
To≈0.1/highest-freq-pole-rad/s

- Phase-lead compensation design procedure:
  - Uses different (more convenient) form of compensator equation D(w):

$$D(z) = K_D \frac{z - z_0}{z - z_P}$$

Note: w_0 and w_P are real-valued

$$D(w) = a_0 \frac{1 + w/w_0}{1 + w/w_P} \implies D(w) = a_0 \frac{1 + a_1 w/a_0}{1 + b_1 w} = \frac{a_0 + a_1 w}{1 + b_1 w}$$

$$where: w = \sigma_w + j\Omega_w; \quad \Omega_w T_0 = 2\tan(\Omega T_0 / 2)$$

Note: a_1 and b_1 are real-valued

- Where:
  - the zero of D(w) using $w_0 = a_0/a_1$
  - the pole of D(w) using $w_P = 1/b_1$
  - And a_0 is the dc gain D(z) at z=1, or D(w) at w=0

Digital Phase-Lead Compensation Procedure

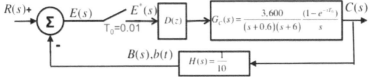

Note: if To is not given, choose

To≈0.1/highest-freq-pole-rad/s

$$D(z) = K_D \frac{z - z_0}{z - z_P}$$

- Phase-lead compensation design procedure:
    1. Take w-transform of open-loop gain $G_{OL}(z)$ with $D(z)=1$ to get $G_{OL}(w)$
    2. Decide upon a phase margin ϕ (default $\phi=45$ degrees)
    3. Define frequency $w_1=j\Omega_{W1}$ where the phase margin is desired
        Default $w_1 = j\Omega_{W1}$ = frequency where $\angle G_{OL}(w_1)$ = -180 degrees
    4. Set gain of D(w) such that $|G_{OL}(w_1)|=1$ (automatically set later), thus

$$\left|G_{OL}(w_1)\right| = \left|D(w_1)\overline{G_C H}(w_1)\right| = 1 \quad \Rightarrow \quad \left|D(w_1)\right| = 1/\left|\overline{G_C H}(w_1)\right|$$

    5. Set phase of D(w) (automatically set later) so $\angle G_{OL}(w_1) = -180+\phi$

$$\angle G_{OL}(w_1) = \angle D(w_1) + \angle \overline{G_C H}(w_1) = -180 + \phi \quad \Rightarrow \quad \angle D(w_1) = -180 + \phi - \angle \overline{G_C H}(w_1) = \theta_D$$

- Where $\angle D(w_1) = \theta_D > 0$

Digital Phase-Lead Compensation Procedure

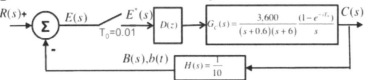

- Phase-lead design procedure, cont'd.:

$$D(z) = K_D \frac{z - z_0}{z - z_P}$$

    6. Set a_0 such that the desired dc gain $G_{OL}(0)$ is obtained
        - Gain determined by other specs such as steady-state error, etc
    7. Solve for a_1 and b_1 and then $w_0 = a_0/a_1$, $w_P = 1/b_1$

$$a_1 = \frac{1 - a_0 \left|\overline{G_C H}(w_1)\right|\cos(\theta_D)}{\text{Im}(w_1)\left|\overline{G_C H}(w_1)\right|\sin(\theta_D)}$$

$$b_1 = \frac{\cos(\theta_D) - a_0 \left|\overline{G_C H}(w_1)\right|}{\text{Im}(w_1)\sin(\theta_D)}$$

Note: a_1 and b_1 are real-valued

Where constraints on D(z) solution include:

$$\angle \overline{G_C H}(w_1) < -180 + \phi, \quad \left|\overline{G_C H}(w_1)\right| < 1/a_0, \quad \cos(\theta_D) > a_0 \left|\overline{G_C H}(w_1)\right|, \quad \text{and} \quad b_1 > 0$$

Digital Phase-Lead Compensation Procedure

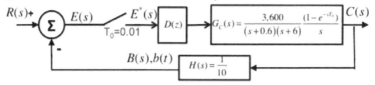

- Phase-lead design procedure, cont'd.:

    8. Convert to z-domain D(z), using w_0, w_P and a_0 from above

$$D(z) = K_D \frac{z - z_0}{z - z_P} = a_0 \frac{1 + w/w_0}{1 + w/w_P}\bigg|_{w = \frac{2}{T_0}\frac{z-1}{z+1}} = \left(a_0 \frac{w_P(1 + w_0 T_0/2)}{w_0(1 + w_P T_0/2)} \right) \frac{z - (1 - w_0 T_0/2)/(1 + w_0 T_0/2)}{z - (1 - w_P T_0/2)/(1 + w_P T_0/2)}$$

where: $w_P > w_0$ for a phase-lead compensator, and

$w = \sigma_w + j\Omega_w$ *and* $z = e^{sT_0}$ *and* $\Omega_w T_0 = 2\tan(\omega/2) = 2\tan(\Omega T_0/2)$

Note: w_0 and w_P are real-valued

    9. And <u>CHECK STABILITY</u> OF $G_{CL}(z)$ since stability is not guaranteed
- Different control system configurations will have different details
- Beware numerical precision pole/zero shifts

Phase-Lead Bode Plots in w-domain

- Phase-lead compensator Bode plot analysis example $D(w) = \frac{1}{\sqrt{10}} \frac{1 + w/1000}{1 + w/10^4}$

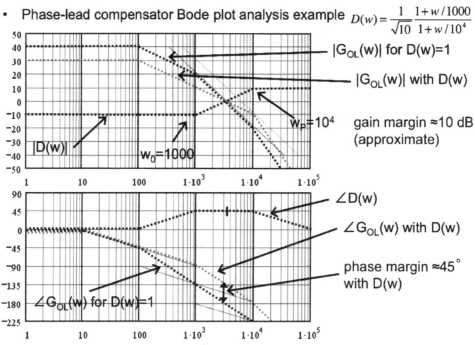

Example Phase-Lead Compensator Design

Example 1: Digital Phase Lead Compensation

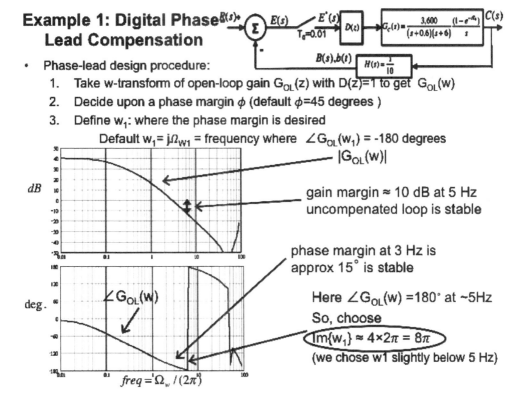

- Phase-lead design procedure:
  1. Take w-transform of open-loop gain $G_{OL}(z)$ with D(z)=1 to get $G_{OL}(w)$
  2. Decide upon a phase margin ϕ (default ϕ=45 degrees)
  3. Define w_1: where the phase margin is desired
     Default $w_1 = j\Omega_{W1}$ = frequency where $\angle G_{OL}(w_1)$ = -180 degrees

$|G_{OL}(w)|$

gain margin ≈ 10 dB at 5 Hz
uncompenated loop is stable

phase margin at 3 Hz is
approx 15° is stable

Here $\angle G_{OL}(w)$ =180° at ~5Hz
So, choose

$Im\{w_1\} \approx 4\times2\pi = 8\pi$

(we chose w1 slightly below 5 Hz)

155

Example 1: Digital Phase-Lead Compensation

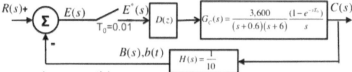

- Phase-lag design procedure, cont'd:
  - ○ From previous slide, $w_1 = j\,8\,\pi$
  4. Ignore (automatically done): ~~Set gain of D(w) such that $|G_{OL}(w_1)|=1$,~~
  5. Ignore (automatically done): ~~Set phase of D(w) so $\angle G_{OL}(w_1) = -180 + \phi$~~
  6. Set a_0 such that the desired dc gain $G_{OL}(0)$ is obtained
     - – Gain determined by other specs such as steady-state error, etc
     - – For present example set $a_0 = 1$ to give approx 1% steady-state error
  7. Solve for a_1 and b_1 and w_0 and w_p

$$\theta_D = \angle D(w_1) = -180 + \phi - \angle \overline{G_C H}(w_1) - 180 + 45 - (-172) = 37^\circ$$

$$a_1 = \frac{1 - a_0 \left|\overline{G_C H}(w_1)\right| \cos(\theta_D)}{\mathrm{Im}(w_1) \left|\overline{G_C H}(w_1)\right| \sin(\theta_D)} = 0.065 \qquad b_1 = \frac{\cos(\theta_D) - a_0 \left|\overline{G_C H}(w_1)\right|}{\mathrm{Im}(w_1)\sin(\theta_D)} = 0.016$$

$$w_0 = a_0 / a_1 = 15.3 \quad and \quad w_p = 1/b_1 = 64.1$$

Example 1: Digital Phase-Lead Compensation

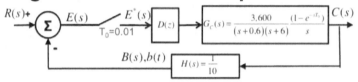

- Phase-lead design procedure, cont'd.:
  8. Convert to z-domain D(z), using w_0, w_p and a_0 from above

$$D(z) = K_D \frac{z - z_0}{z - z_P} = \left(a_0 \frac{w_p\left(1 + w_0 T_0/2\right)}{w_0\left(1 + w_p T_0/2\right)} \right) \frac{z - \left(1 - w_0 T_0/2\right)/\left(1 + w_0 T_0/2\right)}{z - \left(1 - w_p T_0/2\right)/\left(1 + w_p T_0/2\right)}$$

$$D(z) = 3.42 \frac{z - 0.86}{z - 0.52}$$

  9. And CHECK STABILITY OF $G_{CL}(z)$ since stability is not guaranteed
- Different control system configurations will have different details
- Beware numerical precision pole/zero shifts

Example 1: Digital Phase-Lead Compensation

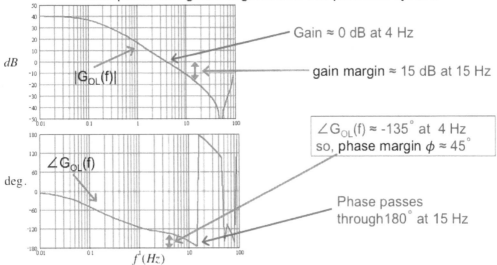

- Phase-lead design procedure, cont'd:
  - o Final Checks:
    - – Check phase and gain margin in new compensated system

Gain ≈ 0 dB at 4 Hz

gain margin ≈ 15 dB at 15 Hz

$\angle G_{OL}(f) \approx -135°$ at 4 Hz
so, phase margin $\phi \approx 45°$

Phase passes
through 180° at 15 Hz

Example 1: Digital Phase-Lead Compensation

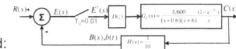

- Phase-lead design procedure, cont'd:
  - o Final Checks:
    - – Check $G_{CL}(z)$ closed loop pulse-transfer function $G_{CL}(z)$

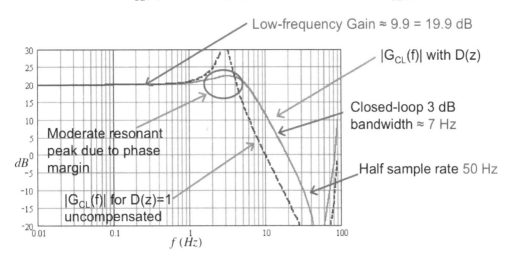

Low-frequency Gain ≈ 9.9 = 19.9 dB

$|G_{CL}(f)|$ with D(z)

Closed-loop 3 dB
bandwidth ≈ 7 Hz

Half sample rate 50 Hz

Moderate resonant
peak due to phase
margin

$|G_{CL}(f)|$ for D(z)=1
uncompensated

$f\ (Hz)$

Example 1: Digital Phase-Lead Compensation

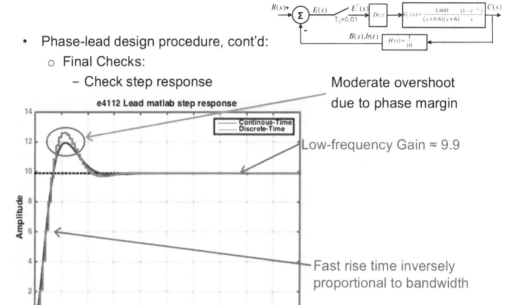

- Phase-lead design procedure, cont'd:
  - ○ Final Checks:
    - – Check step response

Moderate overshoot
due to phase margin

Low-frequency Gain ≈ 9.9

Fast rise time inversely
proportional to bandwidth

Example 1: Digital Phase-Lead Compensation

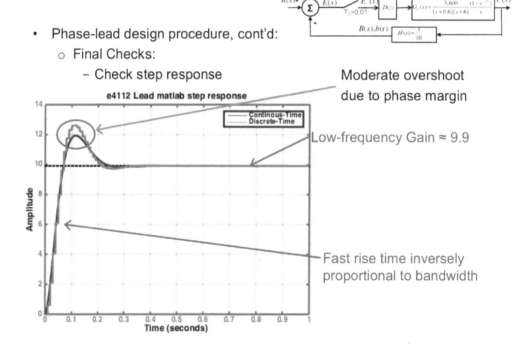

- Phase-lead design procedure, cont'd:
  - ○ Final Checks:
    - – Check step response

Moderate overshoot
due to phase margin

Low-frequency Gain ≈ 9.9

Fast rise time inversely
proportional to bandwidth

Example 1: Digital Phase-Lead Compensation

- Phase-lead responses, cont'd:

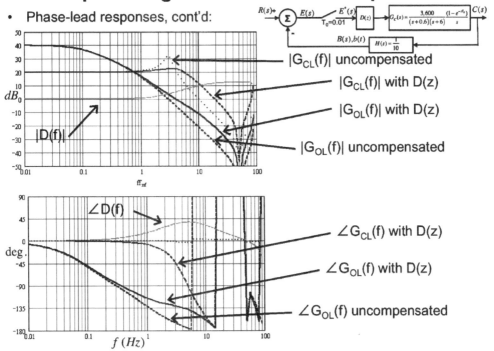

$|G_{CL}(f)|$ uncompensated

$|G_{CL}(f)|$ with D(z)

$|G_{OL}(f)|$ with D(z)

$|G_{OL}(f)|$ uncompensated

$\angle G_{CL}(f)$ with D(z)

$\angle G_{OL}(f)$ with D(z)

$\angle G_{OL}(f)$ uncompensated

Summary: Digital Phase-Lead Compensation

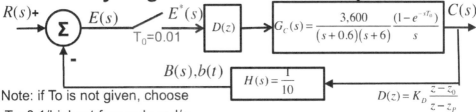

$$D(z) = K_D \frac{z - z_0}{z - z_P}$$

Note: if To is not given, choose
To≈0.1/highest-freq-pole-rad/s

$$D(w) = a_0 \frac{1 + w/w_0}{1 + w/w_P}$$

- Phase-lead compensation design procedure:
  1. Take w-transform of open-loop gain $G_{OL}(z)$ with D(z)=1 to get $G_{OL}(w)$
  2. Decide upon a phase margin ϕ (default ϕ=45 degrees)
  3. Define w_1: where phase margin is desired Default w_1 at $\angle G_{OL}(w_1)$= -180deg.
  4. Ignore (automatically done): ~~Set gain of D(w) such that $|G_{OL}(w_1)|$=1,~~
  5. Ignore (automatically done): ~~Set D(w) so $\angle G_{OL}(w_1)$ = -180+ϕ~~
  6. Set a_0 such that desired dc gain $G_{OL}(0)$; Default set a_0 for ~1% error
  7. Solve for a_1 and b_1 and w_0 and w_P using formulas
  8. Convert to z-domain D(z), using w_0, w_P and a_0 to get D(z)
  9. And <u>CHECK STABILITY</u> OF $G_{CL}(z)$

Advantages/Disadvantages

- Digital phase-lag compensation
  - o Advantages:
    - – Stability margins improved
    - – Low freq. and dc gain can increase (lower error, etc.)
  - o Disadvantages
    - – Reduced bandwidth (could be an advantage)
    - – Low freq zero (near z≈1) can lead to numerical issues (no dc gain)

- Digital phase-lead compensation
  - o Advantages:
    - – Stability margins improved
    - – Increased closed-loop bandwidth (faster rise time, etc.)
    - – Less susceptible to numerical precision issues (poles not near z≈1)
  - o Disadvantages
    - – Increase sensitivity to noise
    - – High gains within loop may lead to large signals
    - – Final design must be checked for stability

Lag-Lead lead Compensator

- Same principles as described for lag and lead compensators
- "Best of both worlds"
- See examples in textbook
- Method:
  - o Design lag section first for dc gain, steady-state error, etc.
  - o Then design lead section
  - o Resulting D(z) has 2 poles and 2 zeroes
  - o D(z) is then just cascade of both designs:

$$D(z) = D_{LAG}(z)D_{LEAD}(z) = K_{DLAG}\frac{z - z_{0LAG}}{z - z_{PLAG}} K_{DLEAD}\frac{z - z_{0LEAD}}{z - z_{PLEAD}}$$

Lag-Lead Bode Plots in w-domain

- Lag-lead compensator response characteristics:

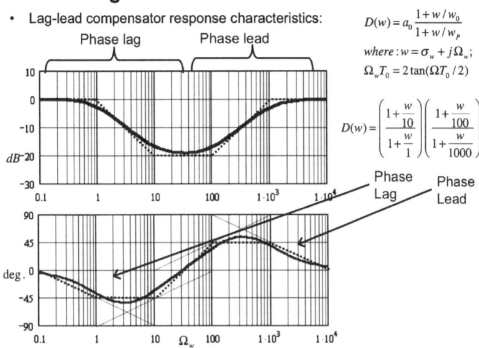

$$D(w) = a_0 \frac{1 + w/w_0}{1 + w/w_P}$$

$$where : w = \sigma_w + j\Omega_w ;$$

$$\Omega_w T_0 = 2\tan(\Omega T_0 / 2)$$

$$D(w) = \left(\frac{1 + \dfrac{w}{10}}{1 + \dfrac{w}{1}} \right) \left(\frac{1 + \dfrac{w}{100}}{1 + \dfrac{w}{1000}} \right)$$

Digital PID Controller Design

Digital PID Compensator

- Lag-lead was "Best of both worlds"
- But, "even better" is PID compensator
- PID – proportional-integral-derivative
- Similar behavior to lag-lead, except integrator at dc
- General form:

$$D(w) = K_P + K_I \frac{1}{w} + K_D \frac{w}{1 + wT_0/2} \approx K_P + K_I \frac{1}{w} + K_D w$$

$$\Rightarrow D(z) = K_P + K_I \frac{T_0}{2} \frac{z+1}{z-1} + K_D \frac{z-1}{T_0 z} \approx K_P + K_I \frac{T_0}{2} \frac{z+1}{z-1} + K_D \frac{2}{T_0} \frac{z-1}{z+1}$$

Preferred approximation of derivative has a left-half plane stable pole at w= -2/To to limit high frequency gain for noise/stability

This approximation of derivative has a pole with unbounded high frequency gain likely to cause instability

- So, implement using the forms on the left
- However, design using forms on the right (easier)

PID Compared to Lag-Lead Bode Plots in w-domain

- Lag-lead compensator response characteristics:

$$D(w) = a_0 \frac{1 + w/w_0}{1 + w/w_P}$$

$$where : w = \sigma_w + j\Omega_w;$$
$$\Omega_w T_0 = 2\tan(\Omega T_0 / 2)$$

$$D(w) = \left(\frac{1 + \dfrac{w}{10}}{1 + \dfrac{w}{1}} \right) \left(\frac{1 + \dfrac{w}{100}}{1 + \dfrac{w}{1000}} \right)$$

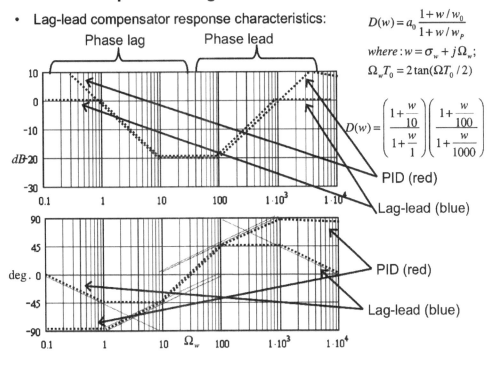

PID (red)

Lag-lead (blue)

PID (red)

Lag-lead (blue)

Digital PID Compensation Overview

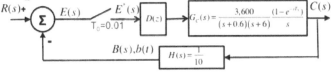

- PID compensation design approach:
  - Small error is induced by the more simple design procedure
  - Uses the more-simple-to design form of compensator equation D(w)
  - However, use the design coefficients in the more stable form for D(z):

$$D(w) = K_P + K_I \frac{1}{w} + K_D w$$

$$\Rightarrow D(z) = K_P + K_I \frac{T_0}{2} \frac{z+1}{z-1} + K_D \frac{z-1}{zT_0}$$

$$where : w = \sigma_w + j\Omega_w; \quad \Omega_w T_0 = 2\tan(\Omega T_0 / 2)$$

- Where the first equation D(w) is used to solve for K_P, K_I, K_D
- And the second equation for D(z) is used for the compensator
  - …even though the D(z) doesn not exactly correspond to D(w)
- Note: text gives more complex exact formulas for proper D(w)

Digital PID Compensation Procedure

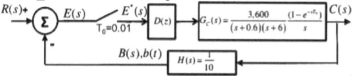

Note: if To is not given, choose
To≈0.1/highest-freq-pole-rad/s

$$D(w) = K_P + K_I \frac{1}{w} + K_D w$$

$$D(z) = K_P + K_I \frac{T_0}{2} \frac{z+1}{z-1} + K_D \frac{z-1}{zT_0}$$

- PID compensation design procedure:
    1. Take w-transform of open-loop gain $G_{OL}(z)$ with $D(z)=1$ to get $G_{OL}(w)$
    2. Decide upon a phase margin ϕ (default ϕ=45 degrees)
    3. Define frequency $w_1 = j\Omega_{W1}$
       Default $w_1 = j\Omega_{W1}$ =half of frequency where $\angle G_{OL}(w)$ = -180+ ϕ for D(z)=1
    4. Set gain of D(w) such that $|G_{OL}(w_1)|$=1 (automatically set later)

$$\left|G_{OL}(w_1)\right| = \left|D(w_1)\overline{G_cH}(w_1)\right| = 1 \;\Rightarrow\; \left|D(w_1)\right| = 1/\left|\overline{G_cH}(w_1)\right|$$

    5. Set phase of D(w) (automatically set later) so $\angle G_{OL}(w_1)$ = -180+ϕ

$$\angle D(w_1) + \angle \overline{G_cH}(w_1) = -180 + \phi \;\Rightarrow\; \angle D(w_1) = -180 + \phi - \angle \overline{G_cH}(w_1) = \theta_D$$

- Where $\angle D(w_1) = \theta_D$ <0 should be negative for a positive K_I

Digital PID Compensation Procedure

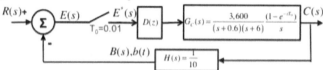

- PID design procedure, cont'd.:
    6. Solve for K_P, K_I, K_D

$$D(w) = K_P + K_I \frac{1}{w} + K_D w$$

$$D(z) = K_P + K_I \frac{T_0}{2} \frac{z+1}{z-1} + K_D \frac{z-1}{zT_0}$$

$$\angle D(w_1) + \angle \overline{G_cH}(w_1) = -180 + \phi \;\Rightarrow\; \boxed{\angle D(w_1) = -180 + \phi - \angle \overline{G_cH}(w_1) = \theta_D}$$

$where:$ $\left|D(w_1)\overline{G_cH}(w_1)\right| = \left|D(w_1)\right|\left|\overline{G_cH}(w_1)\right| = 1$ by definition at phase margin frequency w1
$and:$

$$D(w_1) = K_P + K_I \frac{1}{w_1} + K_D w_1 = K_P + j\left(K_D \Omega_{w1} - \frac{K_I}{\Omega_{w1}}\right) = \left|D(w_1)\right| \angle \theta_D = \left|D(w_1)\right|\left(\cos(\theta_D) + j\sin(\theta_D)\right)$$

equating real and imag. parts gives: $\boxed{K_P = \left|D(w_1)\right|\cos(\theta_D) = \dfrac{\cos(\theta_D)}{\left|\overline{G_cH}(w_1)\right|}}$

$\boxed{and: \quad K_D \Omega_{w1} - \dfrac{K_I}{\Omega_{w1}} = \left|D(w_1)\right|\sin(\theta_D) = \dfrac{\sin(\theta_D)}{\left|\overline{G_cH}(w_1)\right|} \quad where: w_1 = j\Omega_{w1}}$

- Where K_P is defined, but K_D and K_I have some freedom

Digital PID Compensation Procedure

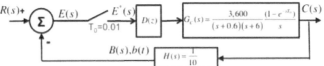

- PID design procedure, cont'd.:

    7. Or, use this default/starting approach:

        – Set K_I by first assuming $K_D=0$ (a PI controller)

            » Make sure θ is negative to assure positive value for K_I

        – Then, set K_P and K_D at frequency $3w_1$, since lead increases bandwidth

$$K_I = \frac{-\Omega_{w1}\sin(\theta_D)}{\overline{|G_cH(w_1)|}} \quad where: w_1 = j\Omega_{w1}, \quad \theta_D = \angle D(w_1) = -180 + \phi - \angle\overline{G_cH}(w_1)$$

$$K_P = \frac{\cos(\theta_{D2})}{\overline{|G_cH(3w_1)|}} \quad where: \theta_{D2} = \angle D(3w_1) = -180 + \phi - \angle\overline{G_cH}(3w_1)$$

$$K_D = \frac{1}{3\Omega_{w1}}\left(\frac{\sin(\theta_{D2})}{\overline{|G_cH(3w_1)|}} + \frac{K_I}{3\Omega_{w1}}\right) \quad where: w_1 = j\Omega_{w1}$$

- Where K_P, K_D and K_I are now all defined   Bumped freq. up to $3w_1$

$$D(w) = K_P + K_I\frac{1}{w} + K_D w$$

$$D(z) = K_P + K_I\frac{T_0}{2}\frac{z+1}{z-1} + K_D\frac{z-1}{zT_0}$$

Digital PID Compensation Procedure

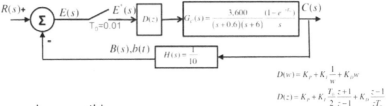

$$D(w) = K_P + K_I\frac{1}{w} + K_D w$$

$$D(z) = K_P + K_I\frac{T_0}{2}\frac{z+1}{z-1} + K_D\frac{z-1}{zT_0}$$

- PID design procedure, cont'd.:

    7. Convert to z-domain D(z), using K_P, K_D and K_I from above

$$D(z) = K_P + K_I\frac{T_0}{2}\frac{z+1}{z-1} + K_D\frac{z-1}{zT_0}$$

$$w = \sigma_w + j\Omega_w \quad and \quad z = e^{sT_0} \quad and \quad \Omega_w T_0 = 2\tan(\omega/2) = 2\tan(\Omega T_0/2)$$

    8. And <u>CHECK STABILITY</u> OF $G_{CL}(z)$ since stability is not guaranteed

- Different control system configurations will have different details
- Beware numerical precision pole/zero shifts

Example Digital PID Controller (Compensator) Design

Example 1: Digital PID Compensation

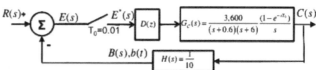

- PID design procedure:
    1. Take w-transform of open-loop gain $G_{OL}(z)$ with $D(z)=1$ to get $G_{OL}(w)$
    2. Decide upon a phase margin ϕ (default ϕ=45 degrees)
    3. Define frequency $w_1 = j\Omega_{W1}$
        Default $w_1 = j\Omega_{W1}$ = half of frequency where $\angle G_{OL}(w) = -180+\phi = -135$ deg

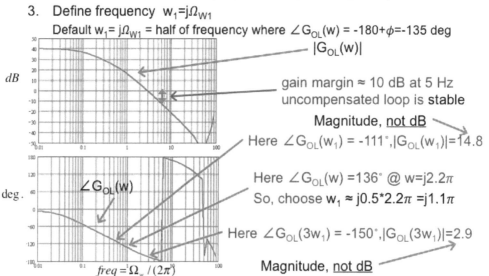

$|G_{OL}(w)|$

gain margin ≈ 10 dB at 5 Hz
uncompensated loop is **stable**

Magnitude, <u>not dB</u>

Here $\angle G_{OL}(w_1) = -111°, |G_{OL}(w_1)|=14.8$

Here $\angle G_{OL}(w) = 136°$ @ $w=j2.2\pi$
So, choose $w_1 \approx j0.5*2.2\pi = j1.1\pi$

Here $\angle G_{OL}(3w_1) = -150°, |G_{OL}(3w_1)|=2.9$

Magnitude, <u>not dB</u>

Example 1: Digital PID Compensation

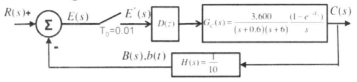

- PID design procedure, cont'd:
  - From previous slide, $w_1 = j1.1\pi$
  4. Ignore (automatically done) ~~Set gain of D(w) such that $|G_{OL}(w_1)|=1$~~
  5. Ignore (automatically done): ~~Set D(w) such that $\angle G_{OL}(w_1) = -180+\phi$~~
  7. Solve for K_P, K_I, K_D, use this default approach:
    - Set K_I such assuming $K_D=0$ (a PI controller)
    - Then, set K_D and Kp at frequency $3w_1$, since lead should increase bandwidth

$$\text{at } w1: \theta_D = \angle D(w_1) = -180 + \phi - \angle \overline{G_c H}(w_1) = -180 + 45 - (-111) = -24°$$

$$\Rightarrow K_I = -\Omega_1 \sin(\theta_D)/\left|\overline{G_c H}(w_1)\right| = -1.1\pi \sin(\theta_D)/14.8 = 0.095$$

$$\text{at } 3w1: \theta_{D2} = \angle D(3w_1) = -180 + \phi - \angle \overline{G_c H}(3w_1) = -180 + 45 - (-150) = 15°$$

$$\Rightarrow K_P = \frac{\cos(\theta_{D2})}{\left|\overline{G_c H}(3w_1)\right|} = \frac{\cos(\theta_{D2})}{2.9} = 0.334 \qquad where: w_1 = j\Omega_{w1}$$

$$and: K_D = \frac{1}{3\Omega_1}\left(\frac{\sin(\theta_{D2})}{\left|\overline{G_c H}(3w_1)\right|} + \frac{K_I}{3\Omega_1}\right) = \frac{1}{3.3\pi}\left(\frac{\sin(\theta_{D2})}{2.9} + \frac{K_I}{3.3\pi}\right) = 0.0093$$

Example 1: Digital PID Compensation

R(s)+ Σ E(s) $T_0=0.01$ E*(s) $D(z)$ $G_c(s) = \dfrac{3.600}{(s+0.6)(s+6)}\dfrac{(1-e^{-sT_0})}{s}$ C(s)

B(s),b(t) $H(s) = \dfrac{1}{10}$

- PID design procedure, cont'd.:
  8. Convert to z-domain D(z), using K_P, K_D and K_I from above

$$D(z) = K_P + K_I \frac{T_0}{2}\frac{z+1}{z-1} + K_D \frac{z-1}{zT_0} = \frac{1.25z^2 - 2.18z + 0.93}{z(z-1)}$$

$$w = \sigma_w + j\Omega_w \quad and \quad z = e^{sT_0} \quad and \quad \Omega_w T_0 = 2\tan(\omega/2) = 2\tan(\Omega T_0/2)$$

  9. And <u>CHECK STABILITY</u> OF $G_{CL}(z)$ since stability is not guaranteed
- Different control system configurations will have different details
- Beware numerical precision pole/zero shifts

Example 1: Digital PID Compensation

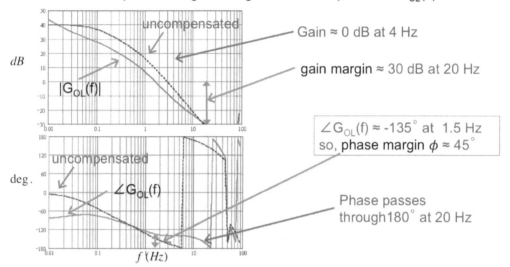

- PID design procedure, cont'd:
  - o Final Checks:
    - – Check phase and gain margin in new compensated $G_{CL}(z)$

Gain ≈ 0 dB at 4 Hz

gain margin ≈ 30 dB at 20 Hz

$\angle G_{OL}(f) \approx -135°$ at 1.5 Hz
so, **phase margin** $\phi \approx 45°$

Phase passes
through 180° at 20 Hz

Example 1: Digital PID Compensation

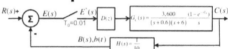

- PID design procedure, cont'd:
  - o Final Checks:
    - – Check $G_{CL}(z)$ closed loop pulse-transfer function $G_{CL}(z)$

Low-frequency Gain ≈ 10 = 20 dB

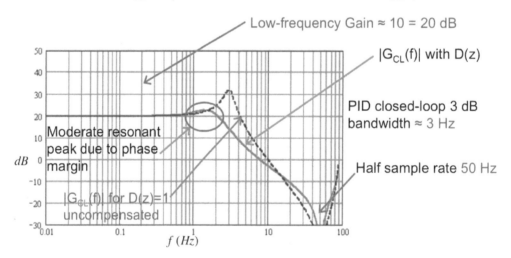

$|G_{CL}(f)|$ with D(z)

PID closed-loop 3 dB
bandwidth ≈ 3 Hz

Half sample rate 50 Hz

Moderate resonant
peak due to phase
margin

$|G_{CL}(f)|$ for D(z)=1
uncompensated

168

Example 1: Digital PID Compensation

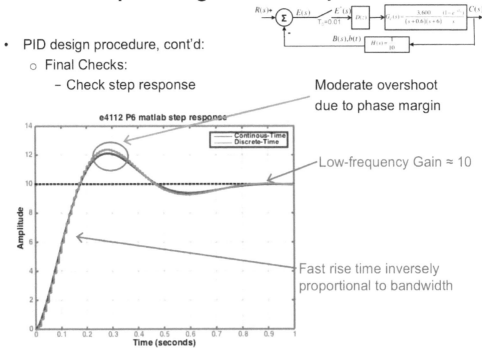

- PID design procedure, cont'd:
  - o Final Checks:
    - – Check step response

Moderate overshoot
due to phase margin

Low-frequency Gain ≈ 10

Fast rise time inversely
proportional to bandwidth

Example 1: Digital PID Compensation

- PIDdesign procedure, cont'd:
  - o Final Checks:
    - – Check poles & margin

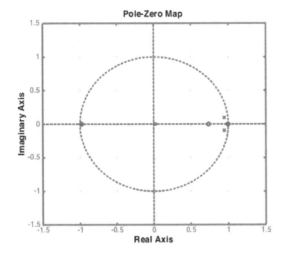

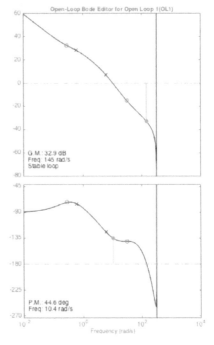

Example 1: Digital PID Compensation

- PID responses, cont'd:

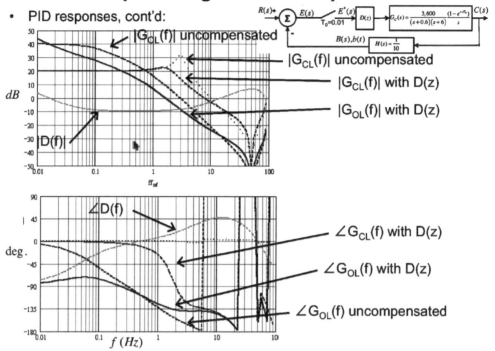

Example 1: Digital PID Compensation

- PID design procedure, cont'd:
  - ○ Final Checks:
    - – Check PID D(z)

PID D(z)

Lead D(z)
(for comparison)

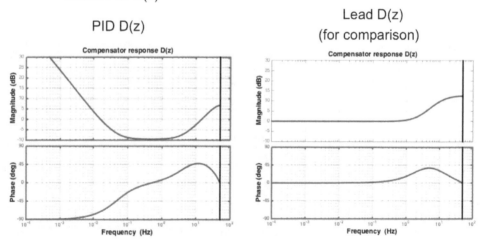

Summary: Digital PID Controller

- Digital PID controller – proportional-integral-derivative
  - Very commonly used
  - PID has similar behavior to lag-lead, except integrator at dc
  - Integrator at low frequency improves steady-state specs
  - PID tuning is a fairly difficult problem
  - PID tuning methods
    - Manual trial/error
    - Automated – mattlab
    - Ziegler–Nichols heuristics
    - Cohen–Coon
    - And others

- General form:

$$D(z) = K_P + K_I \frac{T_0}{2} \frac{z+1}{z-1} + K_D \frac{z-1}{T_0 z}$$

.

11 STATE-VARIABLE CONTROLLERS

This chapter presents the design of state-variable controllers for digital control systems, including the design of observers and controllers.

State-Variable Controllers

Discrete-Time State Variable Methods

- Prior lectures focused on :
  - ○ Analysis using classical SISO (single input single output) approach
  - ○ Uses sampler and starred transform methods
  - ○ Pulse transfer function using z-transform
  - ○ Relationships between Laplace, starred transform, and z-transform
  - ○ Closed-loop and open-loop analyses
  - ○ Stability analyses: gain/phase margin, pole/zero, Routh-Hurwitz, etc
  - ○ Time response: step response, overshoot, rise time, steady-state error
- Next: Discrete-time state-variable approach
  - ○ More modern state-variable approach to system design and analysis
  - ○ Embodies much the same mathematics
  - ○ But more explicitly treats the inner workings (inner state of system)
  - ○ Whereas SISO methods are transfer-function oriented and focus on output and input (not so much the inner workings/state of the system)
  - ○ We will use a hardware approach to derive canonical forms
  - ○ Show state variable form has same transfer function as SISO

State Variable Methods

- Derivation of state-variable approach
  - ○ We will use a hardware approach
  - ○ Starting with classical digital filter block diagram
  - ○ We will modify block diagram into state-variable form
- Canonical forms
  - ○ Two key state canonical forms:
    - – Controllable canonical form
    - – Observable canonical form
- Design with state variables
  - ○ Determining controllability and observability
  - ○ State variable feedback and pole placement
  - ○ Observer design
- Note: there are also continuous time state variable methods

State Variables: A Hardware Derivation

$$y[n] = -a_1 y[n-1] - a_2 y[n-2] ... + b_0 u[n] + b_1 u[n-1] + ...$$

$$Y[z] = -a_1 z^{-1} Y(z) - a_2 z^{-2} Y(z) ... + b_0 U(z) + b_1 z^{-1} U(z) + ...$$

z-transform

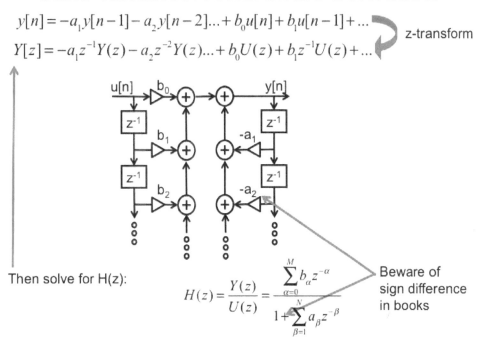

Then solve for H(z):

$$H(z) = \frac{Y(z)}{U(z)} = \frac{\sum_{\alpha=0}^{M} b_\alpha z^{-\alpha}}{1 + \sum_{\beta=1}^{N} a_\beta z^{-\beta}}$$

Beware of sign difference in books

State Variables: A Hardware Derivation

- Start with the same system block diagram as before. $G_{CL}(z)=Y(z)/U(z)$
- We will rearrange the block diagram differently to obtain a new form

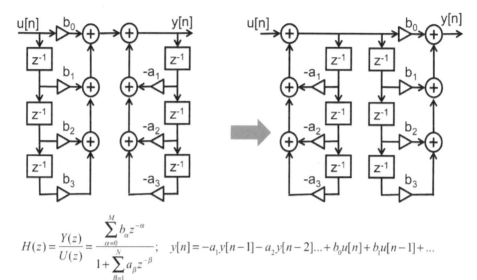

$$H(z)=\frac{Y(z)}{U(z)}=\frac{\displaystyle\sum_{\alpha=0}^{M}b_\alpha z^{-\alpha}}{1+\displaystyle\sum_{\beta=1}^{N}a_\beta z^{-\beta}}; \quad y[n]=-a_1y[n-1]-a_2y[n-2]...+b_0u[n]+b_1u[n-1]+...$$

State Variables: A Hardware Derivation

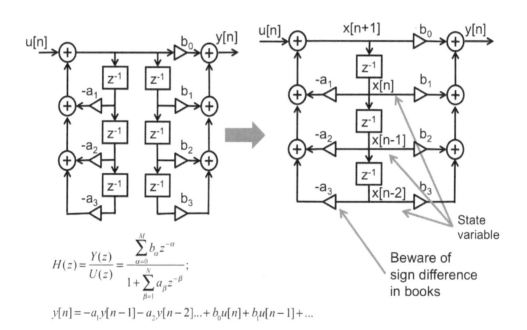

State variable

Beware of sign difference in books

$$H(z)=\frac{Y(z)}{U(z)}=\frac{\displaystyle\sum_{\alpha=0}^{M}b_\alpha z^{-\alpha}}{1+\displaystyle\sum_{\beta=1}^{N}a_\beta z^{-\beta}};$$

$$y[n]=-a_1y[n-1]-a_2y[n-2]...+b_0u[n]+b_1u[n-1]+...$$

State Variables: Controllable Canonical Form
(aka Phase Variable Canonical Form)

- Add state variables: output of each register is a state variable

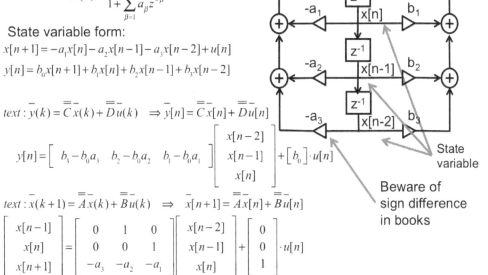

$$H(z) = \frac{Y(z)}{U(z)} = \frac{\displaystyle\sum_{\alpha=0}^{M} b_\alpha z^{-\alpha}}{1 + \displaystyle\sum_{\beta=1}^{N} a_\beta z^{-\beta}}$$

State variable form:

$$x[n+1] = -a_1 x[n] - a_2 x[n-1] - a_3 x[n-2] + u[n]$$

$$y[n] = b_0 x[n+1] + b_1 x[n] + b_2 x[n-1] + b_3 x[n-2]$$

$$text: \overline{y}(k) = \overline{\overline{C}}\,\overline{x}(k) + \overline{\overline{D}}\,\overline{u}(k) \quad \Rightarrow \quad \overline{y}[n] = \overline{\overline{C}}\,\overline{x}[n] + \overline{\overline{D}}\,\overline{u}[n]$$

$$y[n] = \begin{bmatrix} b_3 - b_0 a_3 & b_2 - b_0 a_2 & b_1 - b_0 a_1 \end{bmatrix} \begin{bmatrix} x[n-2] \\ x[n-1] \\ x[n] \end{bmatrix} + \begin{bmatrix} b_0 \end{bmatrix} \cdot u[n]$$

$$text: \overline{x}(k+1) = \overline{\overline{A}}\,\overline{x}(k) + \overline{\overline{B}}\,\overline{u}(k) \quad \Rightarrow \quad \overline{x}[n+1] = \overline{\overline{A}}\,\overline{x}[n] + \overline{\overline{B}}\,\overline{u}[n]$$

$$\begin{bmatrix} x[n-1] \\ x[n] \\ x[n+1] \end{bmatrix} = \begin{bmatrix} 0 & 1 & 0 \\ 0 & 0 & 1 \\ -a_3 & -a_2 & -a_1 \end{bmatrix} \begin{bmatrix} x[n-2] \\ x[n-1] \\ x[n] \end{bmatrix} + \begin{bmatrix} 0 \\ 0 \\ 1 \end{bmatrix} \cdot u[n]$$

State
variable

Beware of
sign difference
in books

State Variable Block Diagram

- The state-variable formulation can be seen in block diagram
- State variables: output of each register is a state variable

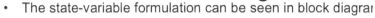

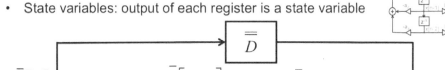

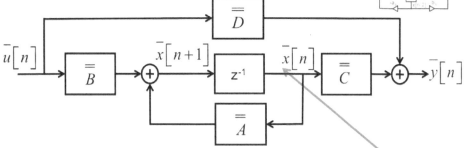

State variable form:

$$\overline{y}[n] = \overline{\overline{C}}\,\overline{x}[n] + \overline{\overline{D}}\,\overline{u}[n]$$

$$\overline{x}[n+1] = \overline{\overline{A}}\,\overline{x}[n] + \overline{\overline{B}}\,\overline{u}[n]$$

$$H(z) = \frac{Y(z)}{U(z)} = \frac{\displaystyle\sum_{\alpha=0}^{M} b_\alpha z^{-\alpha}}{1 + \displaystyle\sum_{\beta=1}^{N} a_\beta z^{-\beta}}$$

State
variable

Example Full Block Diagram

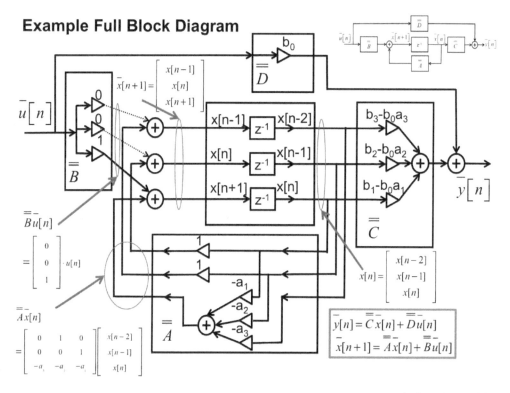

$$\bar{x}[n+1] = \begin{bmatrix} x[n-1] \\ x[n] \\ x[n+1] \end{bmatrix}$$

$$\overline{\overline{B}}\bar{u}[n]$$

$$= \begin{bmatrix} 0 \\ 0 \\ 1 \end{bmatrix} \cdot u[n]$$

$$\overline{\overline{A}}\bar{x}[n]$$

$$= \begin{bmatrix} 0 & 1 & 0 \\ 0 & 0 & 1 \\ -a_3 & -a_2 & -a_1 \end{bmatrix} \begin{bmatrix} x[n-2] \\ x[n-1] \\ x[n] \end{bmatrix}$$

$$x[n] = \begin{bmatrix} x[n-2] \\ x[n-1] \\ x[n] \end{bmatrix}$$

$$\bar{y}[n] = \overline{\overline{C}}\bar{x}[n] + \overline{\overline{D}}\bar{u}[n]$$
$$\bar{x}[n+1] = \overline{\overline{A}}\bar{x}[n] + \overline{\overline{B}}\bar{u}[n]$$

Controllable Mattlab Notation/Form
- The form used in mattlab is:

$$text: \bar{y}(k) = \overline{\overline{C}}\bar{x}(k) + \overline{\overline{D}}\bar{u}(k)$$

$$y[n] = \begin{bmatrix} b_1 - b_0 a_1 & b_2 - b_0 a_2 & b_3 - b_0 a_3 \end{bmatrix} \begin{bmatrix} x[n] \\ x[n-1] \\ x[n-2] \end{bmatrix} + \begin{bmatrix} b_0 \end{bmatrix} \cdot u[n]$$

$$text: \bar{x}(k+1) = \overline{\overline{A}}\bar{x}(k) + \overline{\overline{B}}\bar{u}(k)$$

$$\begin{bmatrix} x[n+1] \\ x[n] \\ x[n-1] \end{bmatrix} = \begin{bmatrix} -a_1 & -a_2 & -a_3 \\ 1 & 0 & 0 \\ 0 & 1 & 0 \end{bmatrix} \begin{bmatrix} x[n] \\ x[n-1] \\ x[n-2] \end{bmatrix} + \begin{bmatrix} 1 \\ 0 \\ 0 \end{bmatrix} \cdot u[n]$$

$$H(z) = \frac{C(z)}{R(z)} = \frac{\sum_{\alpha=0}^{M} b_\alpha z^{-\alpha}}{1 + \sum_{\beta=1}^{N} a_\beta z^{-\beta}}$$

The form used in mattlab is:
bf = [2 3 5 7];
af = [1 -0.11 -0.12 -0.13];
ztf=tf(bf,af,Ts)

$$\frac{2 z^3 + 3 z^2 + 5 z + 7}{z^3 - 0.11 z^2 - 0.12 z - 0.13}$$

A =
0.1100	0.1200	0.1300
1.0000	0	0
0	1.0000	0

C =
| 3.2200 | 5.2400 | 7.2600 |

B =
1
0
0

D =
2

Textbook Notation: Controllable Canonical Form

- Add state variables: output of each register is a state variable

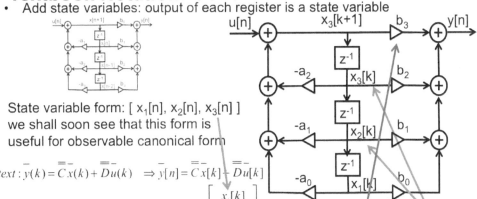

State variable form: [$x_1[n]$, $x_2[n]$, $x_3[n]$]
we shall soon see that this form is
useful for observable canonical form

$$text : \bar{y}(k) = \bar{\bar{C}}\,\bar{x}(k) + \bar{\bar{D}}\,\bar{u}(k) \quad \Rightarrow \bar{y}[n] = \bar{\bar{C}}\,\bar{x}[k] + \bar{\bar{D}}\,\bar{u}[k]$$

State
variable

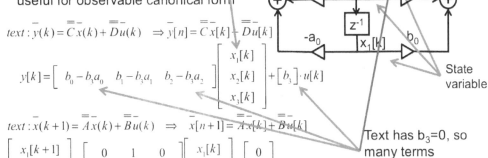

$$y[k] = \begin{bmatrix} b_0 - b_3 a_0 & b_1 - b_3 a_1 & b_2 - b_3 a_2 \end{bmatrix} \begin{bmatrix} x_1[k] \\ x_2[k] \\ x_3[k] \end{bmatrix} + \begin{bmatrix} b_3 \end{bmatrix} \cdot u[k]$$

$$text : \bar{x}(k+1) = \bar{\bar{A}}\,\bar{x}(k) + \bar{\bar{B}}\,\bar{u}(k) \quad \Rightarrow \bar{x}[n+1] = \bar{\bar{A}}\,\bar{x}[k] + \bar{\bar{B}}\,\bar{u}[k]$$

Text has b_3=0, so
many terms
disappear

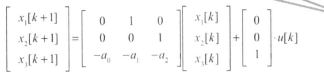

$$\begin{bmatrix} x_1[k+1] \\ x_2[k+1] \\ x_3[k+1] \end{bmatrix} = \begin{bmatrix} 0 & 1 & 0 \\ 0 & 0 & 1 \\ -a_0 & -a_1 & -a_2 \end{bmatrix} \begin{bmatrix} x_1[k] \\ x_2[k] \\ x_3[k] \end{bmatrix} + \begin{bmatrix} 0 \\ 0 \\ 1 \end{bmatrix} \cdot u[k]$$

State-Variable Closed-Loop Response

Solution of Discrete-Time State Variable System

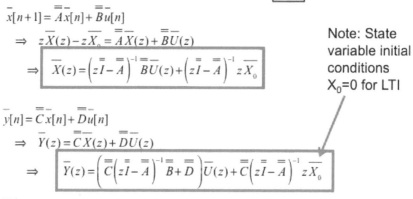

- Solution for state-variable formulation:

$$\bar{x}[n+1] = \bar{\bar{A}}\bar{x}[n] + \bar{\bar{B}}\bar{u}[n]$$

$$\Rightarrow z\bar{X}(z) - z\bar{X}_{\circ} = \bar{\bar{A}}\bar{X}(z) + \bar{\bar{B}}\bar{U}(z)$$

$$\Rightarrow \boxed{\bar{X}(z) = \left(z\bar{\bar{I}} - \bar{\bar{A}}\right)^{-1}\bar{\bar{B}}\bar{U}(z) + \left(z\bar{\bar{I}} - \bar{\bar{A}}\right)^{-1}z\bar{X}_{0}}$$

Note: State variable initial conditions $X_0 = 0$ for LTI

$$\bar{y}[n] = \bar{\bar{C}}\bar{x}[n] + \bar{\bar{D}}\bar{u}[n]$$

$$\Rightarrow \bar{Y}(z) = \bar{\bar{C}}\bar{X}(z) + \bar{\bar{D}}\bar{U}(z)$$

$$\Rightarrow \boxed{\bar{Y}(z) = \left(\bar{\bar{C}}\left(z\bar{\bar{I}} - \bar{\bar{A}}\right)^{-1}\bar{\bar{B}} + \bar{\bar{D}}\right)\bar{U}(z) + \bar{\bar{C}}\left(z\bar{\bar{I}} - \bar{\bar{A}}\right)^{-1}z\bar{X}_{0}}$$

so:

$$\boxed{G_{CL}(z) = \frac{\bar{Y}(z)}{\bar{U}(z)} = \left(\bar{\bar{C}}\left(z\bar{\bar{I}} - \bar{\bar{A}}\right)^{-1}\bar{\bar{B}} + \bar{\bar{D}}\right)} \quad \longleftarrow \text{Transfer Function Matrix}$$

State Transition Matrix (Fundamental Matrix)

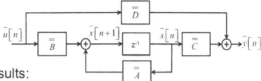

- Taking inverse z-transform of results:

$$\overline{X}(z) = \left(z\overline{\overline{I}} - \overline{\overline{A}} \right)^{-1} \overline{\overline{B}}\,\overline{U}(z) + \left(z\overline{\overline{I}} - \overline{\overline{A}} \right)^{-1} z\overline{X}_0$$

$$so: \quad x[n] = Z^{-1}\left\{ \left(z\overline{\overline{I}} - \overline{\overline{A}} \right)^{-1} \right\} * \overline{\overline{B}}\,\overline{u}[n] + \overline{X}_0\, Z^{-1}\left\{ z\left(z\overline{\overline{I}} - \overline{\overline{A}} \right)^{-1} \right\} = \Phi[n-1] * \overline{\overline{B}}\,\overline{u}[n] + \Phi[n]\overline{X}_0$$

$$\overline{Y}(z) = \left(\overline{\overline{C}}\left(z\overline{\overline{I}} - \overline{\overline{A}} \right)^{-1} \overline{\overline{B}} + \overline{\overline{D}} \right)\overline{U}(z) + \overline{\overline{C}}\left(z\overline{\overline{I}} - \overline{\overline{A}} \right)^{-1} z\overline{X}_0$$

$$so: \quad y[n] = \overline{\overline{C}}\,Z^{-1}\left\{ \left(z\overline{\overline{I}} - \overline{\overline{A}} \right)^{-1} \right\} * \overline{\overline{B}}\,\overline{u}[n] + \overline{\overline{D}}\,\overline{u}[n] + \overline{\overline{C}}\,Z^{-1}\left\{ z\left(z\overline{\overline{I}} - \overline{\overline{A}} \right)^{-1} \right\}\overline{X}_0$$

$$= \overline{\overline{C}}\,\Phi[n-1] * \overline{\overline{B}}\,\overline{u}[n] + \overline{\overline{D}}\,\overline{u}[n] + \overline{\overline{C}}\,\Phi[n]\overline{X}_0 \qquad \text{See text p.74 for } \Phi[n]=A^n$$

$$where \; \Phi[n] = \left(\overline{\overline{A}} \right)^n = Z^{-1}\left\{ z\left(z\overline{\overline{I}} - \overline{\overline{A}} \right)^{-1} \right\} \text{ is the state transition matrix (or fundamental matrix)}$$

State transition matrix $\Phi[n]$ determines time-response of system

$G_{CL}(z)$ and Characteristic Equation of State Variable Form

- To find $G_{CL}(z) = Y(z)/U(z)$ for LTI ($X_0 = 0$):

$$G_{CL}(z) = \frac{Y(z)}{U(z)} = \overline{\overline{C}}\left(z\overline{\overline{I}} - \overline{\overline{A}} \right)^{-1}\overline{\overline{B}} + \overline{\overline{D}}$$

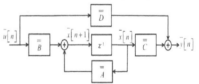

$$\Rightarrow G_{CL}(z) = \frac{\overline{\overline{C}}\,adj\left(z\overline{\overline{I}} - \overline{\overline{A}} \right)\overline{\overline{B}} + \overline{\overline{D}}\left| z\overline{\overline{I}} - \overline{\overline{A}} \right|}{\left| z\overline{\overline{I}} - \overline{\overline{A}} \right|}$$

Zeroes in determinant of denominator determine the poles of closed-loop response

where $adj(A)$ is the adjoint (or adjugate or classical adjoint) of matrix A

- Thus, the zeroes of determinant $| zI - A |$ determine the poles of $G_{CL}(z)$
- And, $| zI - A | = 0$ *is the characteristic equation* of the system (see p. 64)
- Then, partial fraction expansion of the poles sets the impulse response and determines stability/divergence of system

$$G_{CL}(z) = \frac{N(z)}{D(z)} = \frac{z\,Q(z)}{D(z)} = \frac{z\,Q(z)}{(z - p_1)(z - p_2)\circ\circ\circ(z - p_i)} = z\sum_{\alpha=1}^{i}\frac{c_\alpha}{z - p_\alpha}$$

where: $a^n u[n] \Leftrightarrow z/(z-a); \; |z| > |a|$

So: $g_{CL}(z)[n] = Z^{-1}\{G_{CL}(z)\} = Z^{-1}\left\{ \sum_{\alpha=1}^{i}\frac{z c_\alpha}{z - p_\alpha} \right\} = \sum_{\alpha=1}^{i} c_\alpha\left(p_\alpha \right)^n u[n]$

Example: $G_{CL}(z)$ from Controllable State Variable Form

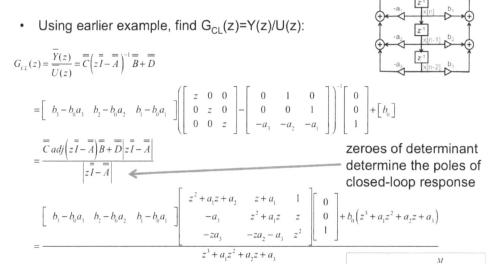

- Using earlier example, find $G_{CL}(z)=Y(z)/U(z)$:

$$G_{CL}(z)=\frac{\overline{Y}(z)}{U(z)}=\overline{\overline{C}}\left(z\overline{\overline{I}}-\overline{\overline{A}}\right)^{-1}\overline{\overline{B}}+\overline{\overline{D}}$$

$$=\begin{bmatrix} b_3-b_0a_3 & b_2-b_0a_2 & b_1-b_0a_1 \end{bmatrix}\left(\begin{bmatrix} z & 0 & 0 \\ 0 & z & 0 \\ 0 & 0 & z \end{bmatrix}-\begin{bmatrix} 0 & 1 & 0 \\ 0 & 0 & 1 \\ -a_3 & -a_2 & -a_1 \end{bmatrix}\right)^{-1}\begin{bmatrix} 0 \\ 0 \\ 1 \end{bmatrix}+\begin{bmatrix} b_0 \end{bmatrix}$$

$$=\frac{\overline{\overline{C}}\,adj\left(z\overline{\overline{I}}-\overline{\overline{A}}\right)\overline{\overline{B}}+\overline{\overline{D}}\left|z\overline{\overline{I}}-\overline{\overline{A}}\right|}{\left|z\overline{\overline{I}}-\overline{\overline{A}}\right|}$$

zeroes of determinant determine the poles of closed-loop response

$$=\frac{\begin{bmatrix} b_3-b_0a_3 & b_2-b_0a_2 & b_1-b_0a_1 \end{bmatrix}\begin{bmatrix} z^2+a_1z+a_2 & z+a_1 & 1 \\ -a_3 & z^2+a_1z & z \\ -za_3 & -za_2-a_3 & z^2 \end{bmatrix}\begin{bmatrix} 0 \\ 0 \\ 1 \end{bmatrix}+b_0\left(z^3+a_1z^2+a_2z+a_3\right)}{z^3+a_1z^2+a_2z+a_3}$$

$$so: G_{CL}(z)=\frac{b_0z^3+b_1z^2+b_2z+b_3}{z^3+a_1z^2+a_2z+a_3}=\frac{b_0+b_1z^{-1}+b_2z^{-2}+b_3z^{-3}}{1+a_1z^{-1}+a_2z^{-2}+a_3z^{-3}}$$

$$H(z)=\frac{Y(z)}{U(z)}=\frac{\sum\limits_{\alpha=0}^{M}b_\alpha z^{-\alpha}}{1+\sum\limits_{\beta=1}^{N}a_\beta z^{-\beta}}$$

- This $G_{CL}(z)$ is the same result as the original $H(z)$

Summary: Discrete-Time State Variable Methods

- We derived state variable approach

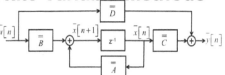

- Discrete-time state-variable methods
  - More modern state-variable approach to system design and analysis
  - Started with known SISO digital filter block diagram with given $H(z)$
  - Rearranged block diagram for controllable state variable form
  - Assigned state variable to each register output
  - Found matrices/vectors A, B, C, D to define state variable
  - General solution includes knowledge of internal register states x[n]
  - Solved for $G_{CL}(z)$ closed-loop gain
  - Show state variable has same $G_{CL}(z)=H(z)$ as original SISO
  - For stability zeroes of determinant | zI - A | must be inside unit circle
  - Stability of system depends on zeroes of determinant | zI - A |, since

$$G_{CL}(z)=\frac{\overline{\overline{C}}\,adj\left(z\overline{\overline{I}}-\overline{\overline{A}}\right)\overline{\overline{B}}+\overline{\overline{D}}\left|z\overline{\overline{I}}-\overline{\overline{A}}\right|}{\left|z\overline{\overline{I}}-\overline{\overline{A}}\right|}$$

Observable Canonical Form

.

State-Variable Observable Canonical Form

- Previous slides: derived state-variable approach
  - o Starting with classical digital filter block diagram
  - o Modified block diagram into state-variable controllable canonical form
  - o Showed state-variable form has same transfer function $G_{CL}(z)$ as SISO

- Next:
  - o Derive state-variable observable canonical form

Observable Canonical Form: Hardware Derivation

- Start with the same system block diagram as before. $G_{CL}(z)=Y(z)/U(z)$
- We will rearrange the block diagram differently to obtain a new form

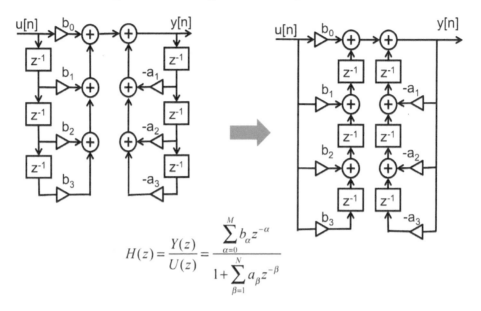

$$H(z) = \frac{Y(z)}{U(z)} = \frac{\displaystyle\sum_{\alpha=0}^{M} b_\alpha z^{-\alpha}}{1 + \displaystyle\sum_{\beta=1}^{N} a_\beta z^{-\beta}}$$

Observable Canonical Form: Hardware Derivation

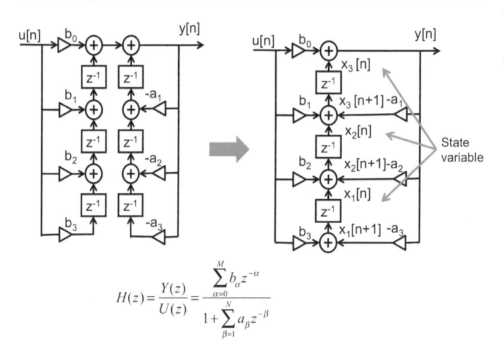

$$H(z) = \frac{Y(z)}{U(z)} = \frac{\displaystyle\sum_{\alpha=0}^{M} b_\alpha z^{-\alpha}}{1 + \displaystyle\sum_{\beta=1}^{N} a_\beta z^{-\beta}}$$

Observable Canonical Form

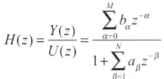

$$H(z) = \frac{Y(z)}{U(z)} = \frac{\sum_{\alpha=0}^{M} b_\alpha z^{-\alpha}}{1 + \sum_{\beta=1}^{N} a_\beta z^{-\beta}}$$

This is state variable observable canonical form:

$$\overline{\overline{A}}_{obs} = \overline{\overline{A}}_{con}^{T}, \quad \overline{\overline{B}}_{obs} = \overline{\overline{C}}_{con}^{T}, \quad \overline{\overline{C}}_{obs} = \overline{\overline{B}}_{con}^{T}, \quad \overline{\overline{D}}_{obs} = \overline{\overline{D}}_{con}$$

and new state vector is $\overline{x}_{obs}[n] = \begin{bmatrix} x_1[n] \\ x_2[n] \\ x_3[n] \end{bmatrix} \neq \overline{x}_{con}[n]$

Note x[n] is not the
same as controllable

$$y[n] = \overline{\overline{C}} x[n] + \overline{\overline{D}} u[n] = \begin{bmatrix} 0 & 0 & 1 \end{bmatrix} \begin{bmatrix} x_1[n] \\ x_2[n] \\ x_3[n] \end{bmatrix} + \begin{bmatrix} b_0 \end{bmatrix} \cdot u[n]$$

$$\overline{x}[n+1] = \begin{bmatrix} x_1[n+1] \\ x_2[n+1] \\ x_3[n+1] \end{bmatrix} = \overline{\overline{A}} x[n] + \overline{\overline{B}} u[n] = \begin{bmatrix} 0 & 0 & -a_3 \\ 1 & 0 & -a_2 \\ 0 & 1 & -a_1 \end{bmatrix} \begin{bmatrix} x_1[n] \\ x_2[n] \\ x_3[n] \end{bmatrix} + \begin{bmatrix} b_3 - b_0 a_3 \\ b_2 - b_0 a_2 \\ b_1 - b_0 a_1 \end{bmatrix} \cdot u[n]$$

Example: $G_{CL}(z)$ from Observable State Variable Form

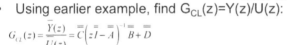

Text has $b_0 = 0$ and
other differences

- Using earlier example, find $G_{CL}(z) = Y(z)/U(z)$:

$$G_{CL}(z) = \frac{Y(z)}{U(z)} = \overline{\overline{C}} \left(z\overline{\overline{I}} - \overline{\overline{A}} \right)^{-1} \overline{\overline{B}} + \overline{\overline{D}}$$

$$= \begin{bmatrix} 0 & 0 & 1 \end{bmatrix} \left(\begin{bmatrix} z & 0 & 0 \\ 0 & z & 0 \\ 0 & 0 & z \end{bmatrix} - \begin{bmatrix} 0 & 0 & -a_3 \\ 1 & 0 & -a_2 \\ 0 & 1 & -a_1 \end{bmatrix} \right)^{-1} \begin{bmatrix} b_3 - b_0 a_3 \\ b_2 - b_0 a_2 \\ b_1 - b_0 a_1 \end{bmatrix} + \begin{bmatrix} b_0 \end{bmatrix}$$

$$= \frac{\overline{\overline{C}} \, adj\left(z\overline{\overline{I}} - \overline{\overline{A}} \right) \overline{\overline{B}} + \overline{\overline{D}} \left| z\overline{\overline{I}} - \overline{\overline{A}} \right|}{\left| z\overline{\overline{I}} - \overline{\overline{A}} \right|}$$

zeroes of determinant
determine the poles of
closed-loop response

$$= \frac{\begin{bmatrix} 0 & 0 & 1 \end{bmatrix} \begin{bmatrix} z^2 + a_1 z + a_2 & -a_3 & -a_3 z \\ z + a_1 & z^2 + a_1 z & -a_2 z - a_3 \\ 1 & z & z^2 \end{bmatrix} \begin{bmatrix} b_3 - b_0 a_3 \\ b_2 - b_0 a_2 \\ b_1 - b_0 a_1 \end{bmatrix} + b_0 \left(z^3 + a_1 z^2 + a_2 z + a_3 \right)}{z^3 + a_1 z^2 + a_2 z + a_3}$$

so: $G_{CL}(z) = \dfrac{b_0 z^3 + b_1 z^2 + b_2 z + b_3}{z^3 + a_1 z^2 + a_2 z + a_3} = \dfrac{b_0 + b_1 z^{-1} + b_2 z^{-2} + b_3 z^{-3}}{1 + a_1 z^{-1} + a_2 z^{-2} + a_3 z^{-3}}$

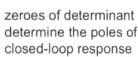

$$H(z) = \frac{Y(z)}{U(z)} = \frac{\sum_{\alpha=0}^{M} b_\alpha z^{-\alpha}}{1 + \sum_{\beta=1}^{N} a_\beta z^{-\beta}}$$

- This $G_{CL}(z)$ is the same result as the original H(z)

State Variable Similarity Transformations

- Consider a similarity transformation of a matrix and corresponding linear change of coordinates:

$\overline{\overline{A}}_s = \overline{\overline{P}}^{-1}\overline{\overline{A}}\overline{\overline{P}}$ where P is any invertible matrix, and also let: $x_s[n] = \overline{\overline{P}}^{-1}\overline{x}[n] \Rightarrow \overline{x}[n] = \overline{\overline{P}}x_s[n]$

then: $\overline{\overline{P}}x_s[n+1] = \overline{\overline{A}}\overline{\overline{P}}x_s[n] + \overline{\overline{B}}u[n] \Rightarrow \overline{\overline{P}}^{-1}\overline{\overline{P}}x_s[n+1] = \overline{\overline{P}}^{-1}\overline{\overline{A}}\overline{\overline{P}}x_s[n] + \overline{\overline{P}}^{-1}\overline{\overline{B}}u[n]$

and: $y[n] = \overline{\overline{C}}\overline{x}[n] + \overline{\overline{D}}u[n] \Rightarrow y[n] = \overline{\overline{C}}\overline{\overline{P}}x_s[n] + \overline{\overline{D}}u[n]$

so: $x_s[n+1] = \overline{\overline{A}}_s x_s[n] + \overline{\overline{P}}^{-1}\overline{\overline{B}}u[n]$ and $y[n] = \overline{\overline{C}}\overline{\overline{P}}x_s[n] + \overline{\overline{D}}u[n]$

- Is a valid state variable form in new coordinates (new state variables x_s)
- Continuing:

$$G_{CL}(z) = \frac{\overline{Y}(z)}{\overline{U}(z)} = \overline{\overline{C}}\overline{\overline{P}}\left(z\overline{\overline{I}} - \overline{\overline{A}}_s\right)^{-1}\overline{\overline{P}}^{-1}\overline{\overline{B}} + \overline{\overline{D}} = \overline{\overline{C}}\left[\overline{\overline{P}}\left(z\overline{\overline{I}} - \overline{\overline{A}}_s\right)\overline{\overline{P}}^{-1}\right]^{-1}\overline{\overline{B}} + \overline{\overline{D}} =$$

$$= \overline{\overline{C}}\left[\overline{\overline{P}}\left(z\overline{\overline{I}} - \overline{\overline{P}}^{-1}\overline{\overline{A}}\overline{\overline{P}}\right)\overline{\overline{P}}^{-1}\right]^{-1}\overline{\overline{B}} + \overline{\overline{D}} = \overline{\overline{C}}\left(z\overline{\overline{P}}\overline{\overline{I}}\overline{\overline{P}}^{-1} - \overline{\overline{P}}\overline{\overline{P}}^{-1}\overline{\overline{A}}\overline{\overline{P}}\overline{\overline{P}}^{-1}\right)^{-1}\overline{\overline{B}} + \overline{\overline{D}}$$

$$\Rightarrow G_{CL}(z) = \overline{\overline{C}}\left(z\overline{\overline{I}} - \overline{\overline{A}}\right)^{-1}\overline{\overline{B}} + \overline{\overline{D}} = \text{SAME AS ORIGINAL!!!}$$

- Thus, $G_{CL}(z)$, poles and zeroes, stability are unchanged!!!

Example: Similarity Transformation

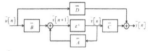

- Consider a similarity transformation that generates permutation of coordinates:

$let\ \overline{\overline{P}} = \begin{bmatrix} 0 & 0 & 1 \\ 0 & 1 & 0 \\ 1 & 0 & 0 \end{bmatrix} so\ \overline{\overline{P}}^{-1} = \begin{bmatrix} 0 & 0 & 1 \\ 0 & 1 & 0 \\ 1 & 0 & 0 \end{bmatrix} and\ \overline{x}_s[n] = \overline{\overline{P}}\overline{x}_s[n] = \begin{bmatrix} 0 & 0 & 1 \\ 0 & 1 & 0 \\ 1 & 0 & 0 \end{bmatrix}\begin{bmatrix} x[n-2] \\ x[n-1] \\ x[n] \end{bmatrix} = \begin{bmatrix} x[n] \\ x[n-1] \\ x[n-2] \end{bmatrix}$

$\overline{x}_s[n+1] = \overline{\overline{A}}_s x_s[n] + \overline{\overline{P}}^{-1}\overline{\overline{B}}u[n] = \begin{bmatrix} 0 & 0 & 1 \\ 0 & 1 & 0 \\ 1 & 0 & 0 \end{bmatrix}\begin{bmatrix} 0 & 1 & 0 \\ 0 & 0 & 1 \\ -a_3 & -a_2 & -a_1 \end{bmatrix}\begin{bmatrix} 0 & 0 & 1 \\ 0 & 1 & 0 \\ 1 & 0 & 0 \end{bmatrix}x_s[n] + \begin{bmatrix} 0 & 0 & 1 \\ 0 & 1 & 0 \\ 1 & 0 & 0 \end{bmatrix}\begin{bmatrix} 0 \\ 0 \\ 1 \end{bmatrix}\cdot u[n]$

$$\Rightarrow \overline{x}_s[n+1] = \begin{bmatrix} -a_1 & -a_2 & -a_3 \\ 1 & 0 & 0 \\ 0 & 1 & 0 \end{bmatrix}\overline{x}_s[n] + \begin{bmatrix} 1 \\ 0 \\ 0 \end{bmatrix}\cdot u[n]$$

and $y[n] = \overline{\overline{C}}\overline{\overline{P}}x_s[n] + \overline{\overline{D}}u[n] = \begin{bmatrix} b_3 - b_0 a_3 & b_2 - b_0 a_2 & b_1 - b_0 a_1 \end{bmatrix}\begin{bmatrix} 0 & 0 & 1 \\ 0 & 1 & 0 \\ 1 & 0 & 0 \end{bmatrix}\overline{x}_s[n] + \begin{bmatrix} b_0 \end{bmatrix}u[n]$

$$\Rightarrow y[n] = \begin{bmatrix} b_1 - b_0 a_1 & b_2 - b_0 a_2 & b_3 - b_0 a_3 \end{bmatrix}\overline{x}_s[n] + \begin{bmatrix} b_0 \end{bmatrix}\cdot u[n]$$

- This controllable form (permutation of coordinates) is the same as the mattlab form noted earlier

Controllability and Observability

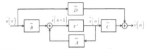

- A system is <u>controllable</u> if there exists some input u[n] that can change the system from any initial vector state x[n] to some new vector state x[n+m], with m finite

- For the case of single-column N-dimensional state vector x[n], and NxN matrix A, the system is controllable if the controllability matrix has rank N:

$$\begin{bmatrix} \overline{B} & \overline{\overline{AB}} & \overline{\overline{A^2 B}} & \cdots & \overline{\overline{A^{N-1} B}} \end{bmatrix}$$

- A system is observable if any initial vector state x[n] can be calculated from a sequence of outputs y[n] through y[n+m] with m finite

- For the case of single-column N-dimensional state vector x[n], and NxN matrix A, the system is observable if the observability matrix has rank N:

V

Principles of State-Variable Control Methods

Adding a State Variable Feedback Controller

- A state-variable feedback controller K is added in this block diagram:

NOTE:
r[n] and u[n]

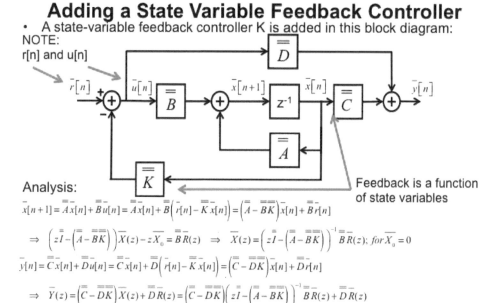

Analysis:

Feedback is a function of state variables

$$\bar{x}[n+1] = \bar{A}\bar{x}[n] + \bar{B}\bar{u}[n] = \bar{A}\bar{x}[n] + \bar{B}\left(\bar{r}[n] - \bar{K}\bar{x}[n]\right) = \left(\bar{A} - \bar{B}\bar{K}\right)\bar{x}[n] + \bar{B}\bar{r}[n]$$

$$\Rightarrow \left(z\bar{I} - \left(\bar{A} - \bar{B}\bar{K}\right)\right)\bar{X}(z) - z\bar{X}_0 = \bar{B}\bar{R}(z) \Rightarrow \bar{X}(z) = \left(z\bar{I} - \left(\bar{A} - \bar{B}\bar{K}\right)\right)^{-1}\bar{B}\bar{R}(z);\ for\ \bar{X}_0 = 0$$

$$\bar{y}[n] = \bar{C}\bar{x}[n] + \bar{D}\bar{u}[n] = \bar{C}\bar{x}[n] + \bar{D}\left(\bar{r}[n] - \bar{K}\bar{x}[n]\right) = \left(\bar{C} - \bar{D}\bar{K}\right)\bar{x}[n] + \bar{D}\bar{r}[n]$$

$$\Rightarrow \bar{Y}(z) = \left(\bar{C} - \bar{D}\bar{K}\right)\bar{X}(z) + \bar{D}\bar{R}(z) = \left(\bar{C} - \bar{D}\bar{K}\right)\left(z\bar{I} - \left(\bar{A} - \bar{B}\bar{K}\right)\right)^{-1}\bar{B}\bar{R}(z) + \bar{D}\bar{R}(z)$$

$$\left(z\bar{I} - \left(\bar{A} - \bar{B}\bar{K}\right)\right)\bar{X}(z) = \bar{B}\bar{R}(z) \Rightarrow \bar{X}(z) = \left(z\bar{I} - \left(\bar{A} - \bar{B}\bar{K}\right)\right)^{-1}\bar{B}\bar{R}(z)$$

Does not assure any closed-loop dc gain

$$so:\ G_{CL}(z) = \frac{\bar{Y}(z)}{R(z)} = \left(\bar{C} - \bar{D}\bar{K}\right)\left(z\bar{I} - \left(\bar{A} - \bar{B}\bar{K}\right)\right)^{-1}\bar{B} + \bar{D}$$

Adding a State Variable Feedback Controller

- For a controllable canonical form of A, B, C, D, this becomes:

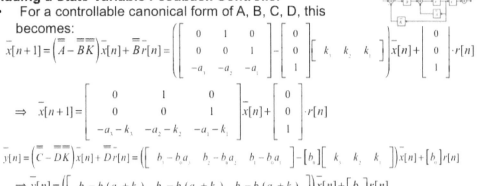

$$x[n+1]=\left(\overline{\overline{A}}-\overline{\overline{B}}\,\overline{\overline{K}}\right)\overline{x}[n]+\overline{\overline{B}}\,\overline{r}[n]=\left(\begin{bmatrix}0 & 1 & 0\\ 0 & 0 & 1\\ -a_1 & -a_2 & -a_1\end{bmatrix}-\begin{bmatrix}0\\0\\1\end{bmatrix}\begin{bmatrix}k_1 & k_2 & k_1\end{bmatrix}\right)\overline{x}[n]+\begin{bmatrix}0\\0\\1\end{bmatrix}\cdot r[n]$$

$$\Rightarrow \overline{x}[n+1]=\begin{bmatrix}0 & 1 & 0\\ 0 & 0 & 1\\ -a_1-k_1 & -a_2-k_2 & -a_1-k_1\end{bmatrix}x[n]+\begin{bmatrix}0\\0\\1\end{bmatrix}\cdot r[n]$$

$$\overline{y}[n]=\left(\overline{\overline{C}}-\overline{\overline{D}}\,\overline{\overline{K}}\right)\overline{x}[n]+\overline{\overline{D}}\,\overline{r}[n]=\left(\begin{bmatrix}b_1-b_0a_1 & b_2-b_0a_2 & b_1-b_0a_1\end{bmatrix}-\begin{bmatrix}b_0\end{bmatrix}\begin{bmatrix}k_1 & k_2 & k_1\end{bmatrix}\right)x[n]+\begin{bmatrix}b_0\end{bmatrix}r[n]$$

$$\Rightarrow \overline{y}[n]=\left(\begin{bmatrix}b_1-b_0(a_1+k_1) & b_2-b_0(a_2+k_2) & b_1-b_0(a_1+k_1)\end{bmatrix}\right)x[n]+\begin{bmatrix}b_0\end{bmatrix}r[n]$$

so: $G_{CL}(z)=\dfrac{Y(z)}{R(z)}=\dfrac{b_0z^3+b_1z^2+b_2z+b_3}{z^3+(a_1+k_1)z^2+(a_2+k_2)z+(a_3+k_3)}$	Feedback can set any denominator coefficients

- Where $G_{CL}(z)$ is by inspection since the form is exactly the same as before, except with a_1 replaced by a_1+k_1, a_2 replaced by a_2+k_2, etc.
- So K can set the poles of $G_{CL}(z)$ anywhere (in theory), since we can set all coefficients of the denominator polynomial! Zeroes of $G_{CL}(z)$ are unaffected
- Practical issues (ridiculously large voltages) may limit this
- Underlying assumption: access to state variables

ModifiedState Variable Feedback Controller

- This form sets closed loop dc gain

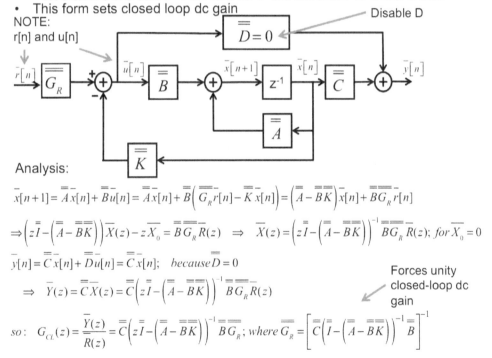

NOTE:
r[n] and u[n]

Analysis:

$$\overline{x}[n+1]=\overline{\overline{A}}\,\overline{x}[n]+\overline{\overline{B}}\,\overline{u}[n]=\overline{\overline{A}}\,\overline{x}[n]+\overline{\overline{B}}\left(\overline{\overline{G}}_R\overline{r}[n]-\overline{\overline{K}}\,\overline{x}[n]\right)=\left(\overline{\overline{A}}-\overline{\overline{B}}\,\overline{\overline{K}}\right)\overline{x}[n]+\overline{\overline{B}}\,\overline{\overline{G}}_R\overline{r}[n]$$

$$\Rightarrow \left(\overline{\overline{z}I}-\left(\overline{\overline{A}}-\overline{\overline{B}}\,\overline{\overline{K}}\right)\right)\overline{X}(z)-z\overline{X}_0=\overline{\overline{B}}\,\overline{\overline{G}}_R R(z) \Rightarrow \overline{X}(z)=\left(\overline{\overline{z}I}-\left(\overline{\overline{A}}-\overline{\overline{B}}\,\overline{\overline{K}}\right)\right)^{-1}\overline{\overline{B}}\,\overline{\overline{G}}_R R(z);\ for\ \overline{X}_0=0$$

$$\overline{y}[n]=\overline{\overline{C}}\,\overline{x}[n]+\overline{\overline{D}}\,\overline{u}[n]=\overline{\overline{C}}\,\overline{x}[n];\quad because\,\overline{\overline{D}}=0$$

$$\Rightarrow \overline{Y}(z)=\overline{\overline{C}}\,\overline{X}(z)=\overline{\overline{C}}\left(\overline{\overline{z}I}-\left(\overline{\overline{A}}-\overline{\overline{B}}\,\overline{\overline{K}}\right)\right)^{-1}\overline{\overline{B}}\,\overline{\overline{G}}_R R(z)$$

Forces unity closed-loop dc gain

$$so:\ G_{CL}(z)=\frac{\overline{Y}(z)}{\overline{R}(z)}=\overline{\overline{C}}\left(\overline{z}I-\left(\overline{\overline{A}}-\overline{\overline{B}}\,\overline{\overline{K}}\right)\right)^{-1}\overline{\overline{B}}\,\overline{\overline{G}}_R;\ where\ \overline{\overline{G}}_R=\left[\overline{\overline{C}}\left(\overline{I}-\left(\overline{\overline{A}}-\overline{\overline{B}}\,\overline{\overline{K}}\right)\right)^{-1}\overline{\overline{B}}\right]^{-1}$$

Adding a State Variable Feedback Controller

- If not in controllable canonical form:

let the desired system be:

$$G_{CL}(z) = \frac{Y(z)}{R(z)} = \frac{Y(z)}{z^N + \alpha_1 z^{N-1} + \cdots + \alpha_{N-1} z + \alpha_N}$$

Note: some texts use different notation

then the desired characteristic equation is:

$$p_{des}(z) = z^N + \alpha_1 z^{N-1} + \cdots + \alpha_{N-1} z + \alpha_N = z^N \left(1 + \alpha_1 z^{-1} + \cdots + \alpha_{N-1} z^{-(N-1)} + \alpha_N z^{-N} \right)$$

define: $p_{des}\left(\overline{\overline{A}} \right) = \overline{\overline{A}}^N + \alpha_1 \overline{\overline{A}}^{N-1} + \cdots + \alpha_{N-1} \overline{\overline{A}} + \alpha_N \overline{\overline{I}}$

then the required K matrix can be computed using Ackermann's formula:

$$\overline{\overline{K}} = \begin{bmatrix} 0 & 0 & 0 & \cdots & 1 \end{bmatrix} \begin{bmatrix} \overline{\overline{B}} & \overline{\overline{A}}\overline{\overline{B}} & \overline{\overline{A}}^2\overline{\overline{B}} & \cdots & \overline{\overline{A}}^{N-1}\overline{\overline{B}} \end{bmatrix}^{-1} p_{des}\left(\overline{\overline{A}} \right)$$

Controllability matrix

Adding a State Variable Estimator

- A state-variable observer is added in this block diagram:

NOTE:
r[n] and u[n]

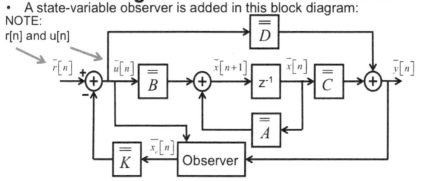

- State variable feedback requires access to state variables
- Often, state variables not accessible, so must estimate from output y[n]
   - Note: if you know the system exactly, you could just simulate it to get x
- Given: A, B, C, D, K, r[n], u[n], y[n], the observer estimates state x[n] as x_e[n]:

given: $\overline{y}[n] = \overline{\overline{C}}\,\overline{x}[n] + \overline{\overline{D}}\,\overline{u}[n]$ *and* $\overline{x}[n+1] = \overline{\overline{A}}\,\overline{x}[n] + \overline{\overline{B}}\,\overline{u}[n]$

form estimates of y[n] and x[n]:

where $\overline{y}_e[n] = \overline{\overline{C}}\,\overline{x}_e[n] + \overline{\overline{D}}\,\overline{u}[n]$ *and*: $\overline{x}_e[n+1] = \overline{\overline{A}}\,\overline{x}_e[n] + \overline{\overline{L}}\left(\overline{y}[n] - \overline{y}_e[n] \right) + \overline{\overline{B}}\,\overline{u}[n]$

190

Estimating State Variables

- State variable feedback requires access to state variables
- Usually state variables not accessible, so must estimate from output y[n]
  - Note: if you know the system exactly, you could just simulate it to get x
- Given: A, B, C, D, K, r[n], u[n], y[n], then estimate state x[n] as $x_e[n]$:

$$given: \quad y[n] = \overline{\overline{C}}\,\overline{x}[n] + \overline{\overline{D}}\,\overline{u}[n] \quad and \quad \overline{x}[n+1] = \overline{\overline{A}}\,\overline{x}[n] + \overline{\overline{B}}\,\overline{u}[n] \quad and \quad \overline{y}_e[n] = \overline{\overline{C}}\,\overline{x}_e[n] + \overline{\overline{D}}\,\overline{u}[n]$$

$$and: \quad \overline{x}_e[n+1] = \overline{\overline{A}}\,\overline{x}_e[n] + \overline{\overline{L}}\left(\overline{y}[n] - \overline{y}_e[n]\right) + \overline{\overline{B}}\,\overline{u}[n]$$

$$= \overline{\overline{A}}\,\overline{x}_e[n] + \overline{\overline{L}}\left(\overline{\overline{C}}\,\overline{x}[n] + \overline{\overline{D}}\,\overline{u}[n] - \overline{\overline{C}}\,\overline{x}_e[n] - \overline{\overline{D}}\,\overline{u}[n]\right) + \overline{\overline{B}}\,\overline{u}[n] = \overline{\overline{A}}\,\overline{x}_e[n] + \overline{\overline{L}}\,\overline{\overline{C}}\left(\overline{x}[n] - \overline{x}_e[n]\right) + \overline{\overline{B}}\,\overline{u}[n]$$

$$then: \quad error = \overline{x}[n+1] - \overline{x}_e[n+1] = \overline{\overline{A}}\left(\overline{x}[n] - \overline{x}_e[n]\right) - \overline{\overline{L}}\,\overline{\overline{C}}\left(\overline{x}[n] - \overline{x}_e[n]\right) = \left(\overline{\overline{A}} - \overline{\overline{L}}\,\overline{\overline{C}}\right)\left(\overline{x}[n] - \overline{x}_e[n]\right)$$

$$\Rightarrow \quad z\left(\overline{X}(z) - \overline{X}_e(z)\right) - z\left(\overline{X} - \overline{X}_e\right)_0 = \left(\overline{\overline{A}} - \overline{\overline{L}}\,\overline{\overline{C}}\right)\left(\overline{X}(z) - \overline{X}_e(z)\right)$$

$$\Rightarrow \quad \frac{\left(\overline{X}(z) - \overline{X}_e(z)\right)}{\left(\overline{X} - \overline{X}_e\right)_0} = \left(z\overline{\overline{I}} - \left(\overline{\overline{A}} - \overline{\overline{L}}\,\overline{\overline{C}}\right)\right)^{-1} = \frac{adj\left(z\overline{\overline{I}} - \left(\overline{\overline{A}} - \overline{\overline{L}}\,\overline{\overline{C}}\right)\right)}{\left|z\overline{\overline{I}} - \left(\overline{\overline{A}} - \overline{\overline{L}}\,\overline{\overline{C}}\right)\right|}$$

Error decays/diminishes so long as poles are inside unit circle

$$and \; so: \left(\overline{x}[n] - \overline{x}_e[n]\right) = Z^{-1}\left\{\left(z\overline{\overline{I}} - \left(\overline{\overline{A}} - \overline{\overline{L}}\,\overline{\overline{C}}\right)\right)^{-1}\right\}\left(\overline{X} - \overline{X}_e\right)_0$$

Where L is to be determined, so as to ensure stable poles

Adding a State Variable Feedback Controller

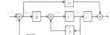
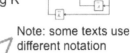

- Note that the form of problem is same as determining K
- So, solution follows similarly:

let the desired system be:

$$\frac{\left(\overline{X}(z) - \overline{X}_e(z)\right)}{\left(\overline{X} - \overline{X}_e\right)_0} = \left(z\overline{\overline{I}} - \left(\overline{\overline{A}} - \overline{\overline{L}}\,\overline{\overline{C}}\right)\right)^{-1} = \frac{adj\left(z\overline{\overline{I}} - \left(\overline{\overline{A}} - \overline{\overline{L}}\,\overline{\overline{C}}\right)\right)}{\left|z\overline{\overline{I}} - \left(\overline{\overline{A}} - \overline{\overline{L}}\,\overline{\overline{C}}\right)\right|} = \frac{Y(z)}{z^N + \alpha_1 z^{N-1} + \cdots + \alpha_{N-1} z + \alpha_N}$$

Note: some texts use different notation

then the desired characteristic equation is:

$$P_{des}(z) = z^N + \alpha_1 z^{N-1} + \cdots + \alpha_{N-1} z + \alpha_N = z^N\left(1 + \alpha_1 z^{-1} + \cdots + \alpha_{N-1} z^{-(N-1)} + \alpha_N z^{-N}\right)$$

$$define: \; p_{des}\left(\overline{\overline{A}}\right) = \overline{\overline{A}}^N + \alpha_1 \overline{\overline{A}}^{N-1} + \cdots + \alpha_{N-1}\overline{\overline{A}} + \alpha_N \overline{\overline{I}}$$

then the required L matrix can be computed using Ackermann's formula:

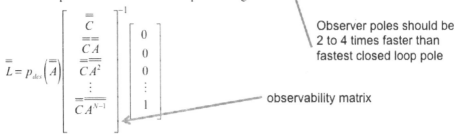

$$\overline{\overline{L}} = p_{des}\left(\overline{\overline{A}}\right)\begin{bmatrix} \overline{\overline{C}} \\ \overline{\overline{C}}\,\overline{\overline{A}} \\ \overline{\overline{C}}\,\overline{\overline{A}}^2 \\ \vdots \\ \overline{\overline{C}}\,\overline{\overline{A}}^{N-1} \end{bmatrix}^{-1}\begin{bmatrix} 0 \\ 0 \\ 0 \\ \vdots \\ 1 \end{bmatrix}$$

Observer poles should be 2 to 4 times faster than fastest closed loop pole

observability matrix

Observer+Controller Block Diagram

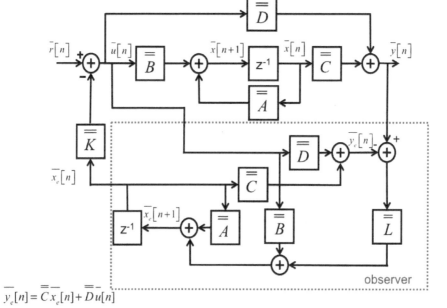

$$\bar{y}_e[n] = \bar{\bar{C}}\bar{x}_e[n] + \bar{\bar{D}}\bar{u}[n]$$

$$\bar{x}_e[n+1] = \bar{\bar{A}}\bar{x}_e[n] + \bar{L}\left(\bar{y}[n] - \bar{y}_e[n]\right) + \bar{\bar{B}}u[n] \quad \Rightarrow \bar{X}_e(z) = z^{-1}\left(\bar{\bar{A}}\bar{X}_e(z) + \bar{L}\left(\bar{Y}(z) - \bar{Y}_e(z)\right) + \bar{\bar{B}}\bar{U}(z)\right)$$

Summary: State Variable Methods

- Discrete-time state-variable methods
    - More modern state-variable approach to system design and analysis
    - Started with known SISO digital filter block diagram with given H(z)
    - Rearranged block diagram for controllable and observable forms
    - Assigned state variable to each register output
    - Found matrices/vectors A, B, C, D to define state variable forms
    - Both controllable and observable forms have same $G_{CL}(z)$
    - Would be impossible to tell the 2 systems apart looking just at $G_{CL}(z)$
    - So, both controllable and observable forms have same poles/zeroes
    - So, both controllable and observable forms have same stability
    - Observable form: Because output y[n] arises at the end of a chain of the state-registers, the output is affected by all of the state registers, and thus the effect of every state register should be observable in y[n]
    - Controllable form: Because input u[n] appears at the input of the chain of state registers, it should be possible to change the state of all of the state variables by some change of the input u[n]
    - **State feedback theoretically allows arbitrary pole placement**

192

System Identification Basics

System Identification

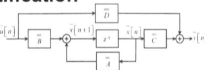

- Last time: Discrete-time state-variable approach
  - o More modern state-variable approach to system design and analysis
  - o It more explicitly treats the inner workings (inner state of system)
  - o Whereas SISO methods are transfer-function oriented and focus on output and input (not so much the inner workings/state of the system)
  - o Used a hardware approach to derive canonical forms
  - o Showed state variable form has same transfer function as SISO

- Next:
  - o Above presumes state-variable model is known: A, B, C, D
  - o How do we determine the model for discrete-time systems?
  - o Today we explore ARMA model estimation
    - – Also, many other methods: see text

ARMA Model

- Recall model of digital filter:

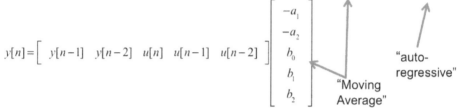

$$H(z) = \frac{Y(z)}{U(z)} = \frac{\sum_{\alpha=0}^{M} b_\alpha z^{-\alpha}}{1 + \sum_{\beta=1}^{N} a_\beta z^{-\beta}};$$

$$y[n] = -a_1 y[n-1] - a_2 y[n-2]... + b_0 u[n] + b_1 u[n-1] + ...$$

$$Y[z] = -a_1 z^{-1} Y(z) - a_2 z^{-2} Y(z)... + b_0 U(z) + b_1 z^{-1} U(z) + ...$$

- Can also be rewritten as:
- (second-order model example)

$$y[n] = \begin{bmatrix} y[n-1] & y[n-2] & u[n] & u[n-1] & u[n-2] \end{bmatrix} \begin{bmatrix} -a_1 \\ -a_2 \\ b_0 \\ b_1 \\ b_2 \end{bmatrix}$$

"Moving Average"   "auto-regressive"

- This is an ARMA model (autoregressive moving average)
- NOTE: order of model presumed known beforehand

ARMA Model

- Furthermore (second-order model example):

$$\begin{bmatrix} y[n] \\ y[n+1] \\ y[n+2] \\ y[n+3] \\ \vdots \\ y[n+N-1] \end{bmatrix} \overset{?}{=} \begin{bmatrix} y[n-1] & y[n-2] & u[n] & u[n-1] & u[n-2] \\ y[n] & y[n-1] & u[n+1] & u[n] & u[n-1] \\ y[n+1] & y[n] & u[n+2] & u[n+1] & u[n] \\ y[n+2] & y[n+1] & u[n+3] & u[n+2] & u[n+1] \\ \vdots & \vdots & \vdots & \vdots & \vdots \\ y[n+N-2] & y[n+N-3] & u[n+N-1] & u[n+N-2] & u[n+N-3] \end{bmatrix} \begin{bmatrix} -a_1 \\ -a_2 \\ b_0 \\ b_1 \\ b_2 \end{bmatrix}$$

- Can also be rewritten as:

$$\bar{Y}[n] \overset{?}{=} \bar{\bar{F}}[n]\bar{\theta} \quad or \quad \bar{Y}[n] = \bar{\bar{F}}[n]\bar{\theta} + \bar{e}[n]$$

- Where least-square error e[n] is minimized using left pseudoinverse:

$$\bar{\theta} = \begin{bmatrix} -a_1 \\ -a_2 \\ b_0 \\ b_1 \\ b_2 \end{bmatrix} = \left(\bar{\bar{F}}^T[n]\bar{\bar{F}}[n] \right)^{-1} \bar{\bar{F}}^T[n]\bar{Y}[n]$$

- This is least-squares estimate of the filter parameters
- Note: text notation differs from handouts

ARMA Model

- Once filter parameters are known, any form of realization may be chosen

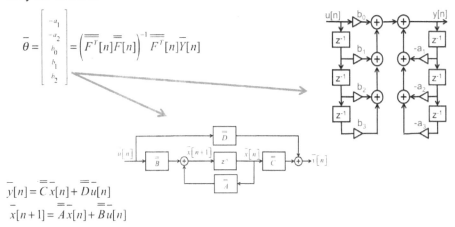

$$\bar{\theta} = \begin{bmatrix} -a_1 \\ -a_2 \\ b_0 \\ b_1 \\ b_2 \end{bmatrix} = \left(\overline{\overline{F}}^T[n]\overline{\overline{F}}[n]\right)^{-1} \overline{\overline{F}}^T[n]\bar{Y}[n]$$

$$\bar{y}[n] = \overline{\overline{C}}\,\bar{x}[n] + \overline{\overline{D}}\,\bar{u}[n]$$
$$\bar{x}[n+1] = \overline{\overline{A}}\,\bar{x}[n] + \overline{\overline{B}}\,\bar{u}[n]$$

- Free to choose form of system:
  - o controllable canonical,
  - o observable canonical,
  - o etc.

Example: ARMA Model

- Furthermore (second-order model example),
- Ydat is impulse response first few points:

$$\frac{z^2 + 2z + 3}{\left(z + \frac{1}{2}\right)\left(z + \frac{1}{3}\right)}$$

$$F = \begin{vmatrix} 0 & 0 & 1 & 0 & 0 \\ 1 & 0 & 0 & 1 & 0 \\ 1.167 & 1 & 0 & 0 & 1 \\ 1.861 & 1.167 & 0 & 0 & 0 \\ -1.745 & 1.861 & 0 & 0 & 0 \\ 1.144 & -1.745 & 0 & 0 & 0 \end{vmatrix} , \quad Yn = \begin{vmatrix} 1 \\ 1.167 \\ 1.861 \\ -1.745 \\ 1.144 \\ -0.663 \end{vmatrix} ,$$

- Solving:

$$\bar{\theta} = \begin{bmatrix} -a_1 \\ -a_2 \\ b_0 \\ b_1 \\ b_2 \end{bmatrix} = \left(\overline{\overline{F}}^T[n]\overline{\overline{F}}[n]\right)^{-1} \overline{\overline{F}}^T[n]\bar{Y}[n]$$

$$\theta = \left[\left(F^T F\right)^{-1} F^T Yn\right] \longrightarrow \theta = \begin{vmatrix} -0.833 \\ -0.167 \\ 1 \\ 2 \\ 3 \end{vmatrix} ,$$

- This example recovers numerator and denominator coefficients of H(z)

12 APPLICATIONS

This chapter provides discussion of applications of digital control system theory to the design of digital impedances and the design of phase-locked loops.

Digital Impedance

Digital Impedance Design

- A new area of research is in *digital impedance design*
- The notion of digital impedance design is to replace some continuous time impedance by a digital equivalent
- In simplified terms the basic concept is:

$$Z_{in}(s) = \frac{V(s)}{I(s)} \approx \left. \frac{V(z)}{I(z)} \right|_{z=e^{sT}}$$

- Allows tunable/adjustable/adaptive impedances
- Make "normal" impedances: resistance, capacitance, etc.
- Most useful for "exotic" impedances:
    - Negative capacitance
    - Negative inductance

What is Digital Impedance?

Simple Example: Digital Resistor

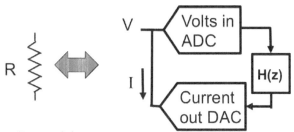

- Measure voltage V
- Set current I
- Let H(z) = 1/R
- So, DAC current: I = V/R

... yields world's most expensive resistor!
... but is tunable

Digital Impedance: Theory

- Measure v(t)   ADC: v[n] = v(nT)
- Calculate i[n]   H(z): i[n] = v[n]*h[n]
- Set i(t)   DAC: i(nT) = i[n]

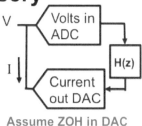

Assume ZOH in DAC

$$I(s) = V^*(s)\, H(z) \left.\frac{(1-z^{-1})}{s}\right|_{z=e^{sT}}$$

ZOH

$$V^*(s) = \sum v(nT)e^{-nsT}$$
$$= \sum V(s - n\omega_0)/T$$

(Starred Transform)

Input impedance is then:

$$Z_{in}(s) = \frac{V(s)}{I(s)} \approx \left.\frac{sT}{\left[\left(1-z^{-1}\right)H(z)\right]}\right|_{z=e^{sT}}$$

Here let V*(s) ≈V(s) at low frequency, for properly sampled signals

Digital Capacitor Example

Capacitor:

$$i(t) = C\frac{dv(t)}{dt} \approx C\frac{v[n]-v[n-1]}{T}$$

Taking z-transform:

$$I(z) = C(1-z^{-1})V(z)/T = H(z)V(z)$$

Yields the transfer function H(z):

$$H_C(z) = C(1-z^{-1})/T$$

The input impedance is then:

$$Z_C(s) = \frac{sT}{\left[(1-z^{-1})H(z)\right]}\bigg|_{z=e^{sT}} = \frac{sT^2}{C(1-z^{-1})^2}\bigg|_{z=e^{sT}} \approx \frac{1}{sC}$$

Note: C can be positive or negative

New Thevenin Approach

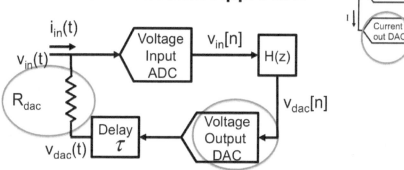

- Voltage DAC plus R_{dac} is a Thevenin source
- Could do simple *Thevenin-to-Norton* transformation
- Better approach: incorporate R_{dac} into H(z) design
- Time-delay latencies are also modeled
- Latency caused by ADC conversion and computation time

Thevenin Approach: General Theory

- Measure $v(t)$ ADC: $v[n] = v(nT)$
- Calculate $v_{dac}[n]$ $H(z)$: $v_{dac}[n] = v[n]*h[n]$
- Set $v_{dac}(t)$ DAC: $v_{dac}(nT) = v_{dac}[n]$

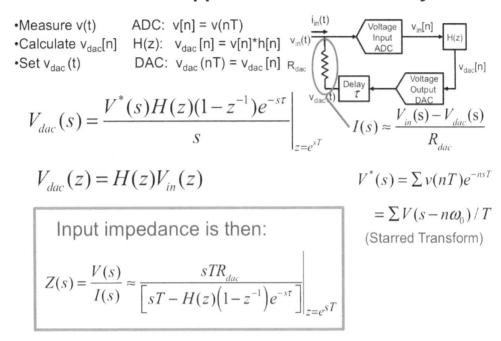

$$V_{dac}(s) = \left. \frac{V^*(s)H(z)(1-z^{-1})e^{-s\tau}}{s} \right|_{z=e^{sT}}$$

$$I(s) \approx \frac{V_{in}(s) - V_{dac}(s)}{R_{dac}}$$

$$V_{dac}(z) = H(z)V_{in}(z)$$

$$V^*(s) = \sum v(nT)e^{-nsT}$$

$$= \sum V(s-n\omega_0)/T$$

(Starred Transform)

Input impedance is then:

$$Z(s) = \frac{V(s)}{I(s)} \approx \left. \frac{sTR_{dac}}{\left[sT - H(z)\left(1-z^{-1}\right)e^{-s\tau} \right]} \right|_{z=e^{sT}}$$

Thevenin Form Digital RC circuit

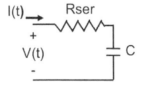

Thevenin Form Digital RC Circuit: Theory

Series RC circuit:

$$v(t) = i(t)R_{ser} + \int \frac{i(t)dt}{C}$$

Taking the derivative:

$$\frac{dv_{in}(t)}{dt} = R_{ser}\frac{di_{in}(t)}{dt} + \frac{i(t)}{C}$$

Discrete-time approximations:

$$\frac{dv_{in}(t)}{dt} \approx \frac{v_{in}[n] - v_{in}[n-1]}{T} \qquad i_{in}[n] \approx \frac{v_{in}[n] - v_{dac}[n]}{R_{dac}}$$

Yields:

$$v_{dac}[n](R_{ser}C + T) = v_{in}[n](R_{ser}C - R_{dac}C + T)$$

$$+ v_{in}[n-1](R_{dac}C - R_{ser}C) + v_{dac}[n-1]R_{ser}C$$

Thevenin Form Digital RC Circuit: Theory

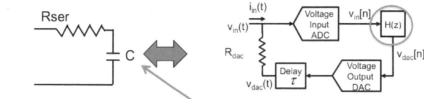

From previous slide:

$$v_{dac}[n](R_{ser}C + T) = v_{in}[n](R_{ser}C - R_{dac}C + T)$$

Note: C can be positive or negative

$$+ v_{in}[n-1](R_{dac}C - R_{ser}C) + v_{dac}[n-1]R_{ser}C$$

Taking the z transform to solve for H(z):

$$H_{RC}(z) = \frac{V_{dac}(z)}{V_{in}(z)} = \frac{(R_{ser}C - R_{dac}C + T)z + (R_{dac}C - R_{ser}C)}{(R_{ser}C + T)z - R_{ser}C}$$

Stability Analysis with Source Impedance

- Add **External Source** to Analyze Stability

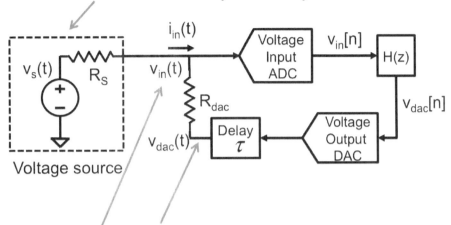

- Here: $v_{in}(t) = v_{dac}(t - \tau) + R_{dac} (v_s(t) - v_{dac}(t - \tau)) / (R_s + R_{dac})$

Stability Analysis for $\tau = 0$

Then:

$$V_{in}^*(s) = \frac{V_{dac}^*(s)}{H(z)}\Big|_{z=e^{sT}} = V_{dac}^*(s)e^{-s\tau} + \frac{R_{dac}}{R_s + R_{dac}}\left[V_s^*(s) - V_{dac}^*(s)e^{-s\tau} \right]$$

Solving for the transfer function, and for simplicity, let $\tau = 0$:

$$\frac{V_{dac}^*(s)}{V_s^*(s)} = \frac{R_{dac} H(z)}{R_s + R_{dac} - R_s H(z)}\Big|_{z=e^{sT}}$$

$$\text{where } V_{dac}^*(s) = \left(V_{in}^*(s) H(z)\big|_{z=e^{sT}} \left(1 - e^{-sT_0}\right)/s \right)^* = V_{in}^*(s) H(z)\big|_{z=e^{sT}}$$

Stable if poles of closed-loop *pulse transfer function* $G_{CL}(z)$ are inside the unit circle:

$$G_{CL}(z) = \frac{V_{dac}(z)}{V_s(z)} = \frac{R_{dac} H(z)}{R_s + R_{dac} - R_s H(z)}$$

Stability Analysis for $\tau = 0$

For the digital RC :

$$H_{RC}(z) = \frac{(R_{ser}C - R_{dac}C + T)z + (R_{dac}C - R_{ser}C)}{(R_{ser}C + T)z - R_{ser}C}$$

The overall transfer function for $\tau = 0$ is then:

$$G_{CLdelay0}(z) = \frac{R_{dac}H(z)}{(R_s + R_{dac}) - R_s H(z)}$$

$$= \frac{(R_{ser}C - R_{dac}C + T)z + (R_{dac}C - R_{ser}C)}{z(R_{ser}C + R_sC + T) - R_{ser}C - R_sC}$$

Note: C can be positive or negative

$$pole: z = \frac{R_{ser}C + R_sC}{R_{ser}C + R_sC + T} = \frac{1}{1 + \dfrac{T}{R_{ser}C + R_sC}}$$

> Which is stable for poles inside unit circle:
>
> $$\frac{T}{R_{ser}C + R_sC} > 0 \quad or \quad \frac{T}{R_{ser}C + R_sC} < -2$$

(for real R_S, T, and C)

Stability Analysis for τ = T, R_{ser}=0

- As a second simplification to gain insight:
  - For simplicity, let τ = T = 1 clock period, and R_{ser}=0,
  - The transfer function is:

$$G_{CLdelay1}(z) = \frac{R_{dac}z^{-1}H(z)}{R_s + R_{dac} - R_s z^{-1}H(z)} = \frac{R_{dac}H(z)}{z(R_s + R_{dac}) - R_s H(z)}$$

$$= \frac{R_{dac}(T - R_{dac}C + T)z + R_{dac}^2 C}{(TR_s + TR_{dac})z^2 + (R_s R_{dac}C - R_s T)z - R_s(R_{dac}C)}$$

Note: C can be positive or negative

- System is stable if the poles inside the unit circle
- However, now more complicated quadratic form
- Two poles
- Best to study using root locus

Stability Analysis Details for $\tau = T$

Details for the case with $\tau = T$:

$$V_{in}^*(s) = \frac{V_{dac}^*(s)}{H(z)\big|_{z=e^{sT}}} = V_{dac}^*(s)e^{-s\tau} + \frac{R_{dac}}{R_s + R_{dac}}\left[V_s^*(s) - V_{dac}^*(s)e^{-s\tau}\right]$$

$$V_{dac}^*(s) = V_{dac}^*(s)e^{-s\tau}H(z)\big|_{z=e^{sT}} + \frac{R_{dac}H(z)\big|_{z=e^{sT}}}{R_s + R_{dac}}\left[V_s^*(s) - V_{dac}^*(s)e^{-s\tau}\right]$$

$where\ V_{dac}^*(s) = \left(V_{in}^*(s)H(z)\big|_{z=e^{sT}}\left(1 - e^{-sT_o}\right)/s\right)^* = V_{in}^*(s)H(z)\big|_{z=e^{sT}}$

$$V_{dac}^*(s)\left\{1 - z^{-1}H(z) + \frac{R_{dac}}{R_s + R_{dac}}z^{-1}H(z)\right\} = V_{dac}^*(s)\left\{1 + \frac{-R_s}{R_s + R_{dac}}z^{-1}H(z)\right\} = \frac{R_{dac}}{R_s + R_{dac}}z^{-1}H(z)V_s^*(s)$$

$$V_{dac}^*(s)\left\{R_s + R_{dac} - R_s z^{-1}H(z)\right\} = R_{dac}z^{-1}H(z)V_s^*(s)$$

$$\frac{V_{dac}^*(s)}{V_s^*(s)} = \frac{R_{dac}z^{-1}H(z)}{R_s + R_{dac} - R_s z^{-1}H(z)}\bigg|_{z=e^{sT}}$$

Finally:

$$G_{CL}(z) = \frac{V_{dac}(z)}{V_s(z)} = \frac{R_{dac}z^{-1}H(z)}{R_s + R_{dac} - R_s z^{-1}H(z)}; \quad \text{for 1-clock-cycle latency}$$

Examples
Digital Negative and Positive Capacitances

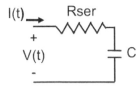

Easy Analog Negative Capacitor

- Before proceeding
  - Quick overview of negative capacitance

- Negative resistor: Zin = Vin/Iin = -R

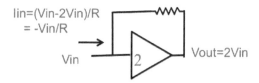

- Negative capacitor: Zin = Vin/Iin = -1/(sC)

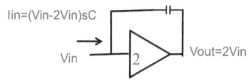

Why Non-Foster Circuits?

- Of particular interest:
  - Negative capacitors
  - Negative inductors

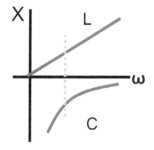

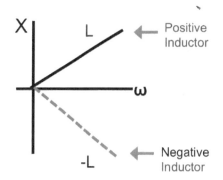

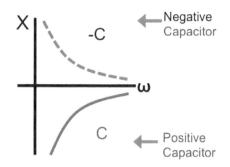

Example Negative Cap.: τ=T=1μs, R$_{ser}$=100, C=-5nF, R$_{dac}$=1000

- As first design step:
    - Use matthcad to do root locus to find stable R$_{ser}$:

root locus as Rser is varied ifrom -RserMax to RserMax in nR steps

Rsa = 50

Ca = $-5 \cdot 10^{-9}$

Ta = $1 \cdot 10^{-6}$

Rdaca = $1 \cdot 10^{3}$

RserMin = -340

RserMax = 160

Rsteps = 32

nR = 10

RLatency = 1

Rser=-160

Rser=-340

ImagZeroes
ooo
ImagPoles
xxx
ImagUnitCirc

RealZeroes RealPoles RealUnitCirc

- Above R$_{ser}$=100 seems ~best near unit circle

Example Negative Cap.: τ=T=1μs, R$_{ser}$=100, C=-5nF, R$_{dac}$=1000

- As next step:
    - Use matthcad to determine Zin for R$_{ser}$:

Rsa = 50   Rsera = 100   Ca = -5×10^{-9} Ta = 1×10^{-6} Rdaca = 1×10^{3}

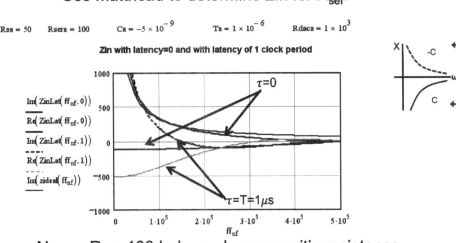

Zin with latency=0 and with latency of 1 clock period

$\text{Im}(\text{ZinLat}(\text{ff}_{nf}, 0))$

$\text{Re}(\text{ZinLat}(\text{ff}_{nf}, 0))$

$\text{Im}(\text{ZinLat}(\text{ff}_{nf}, 1))$

$\text{Re}(\text{ZinLat}(\text{ff}_{nf}, 1))$

$\text{Im}(\text{zideal}(\text{ff}_{nf}))$

τ=0

τ=T=1μs

ff$_{nf}$

- Above R$_{ser}$=100 helps reduce parasitic resistance

Example Negative Cap.: $\tau=T=1\mu s$, $R_{ser}=100$, $C=-5nF$, $R_{dac}=1000$

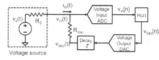

- ## As next step:
    - ## Use matthcad to determine Zin for R_{ser}:

$Rsa = 50$ $Rsera = 100$ $Ca = -5 \times 10^{-9}$ $Ta = 1 \times 10^{-6}$ $Rdaca = 1 \times 10^3$

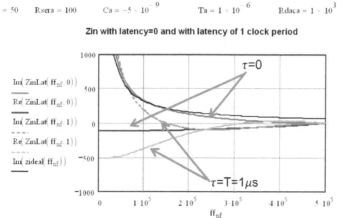

- ## Above $R_{ser}=100$ helps reduce parasitic resistance

Example Negative Cap.: $\tau=T=1\mu s$, $R_{ser}=100$, $C=-5nF$, $R_{dac}=1000$

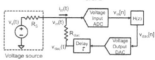

- ## As next step:
    - ## Use matthcad to determine Zin for R_{ser}:

$Rsa = 50$ $Rsera = 100$ $Ca = -5 \times 10^{-9}$ $Ta = 1 \times 10^{-6}$ $Rdaca = 1 \times 10^3$

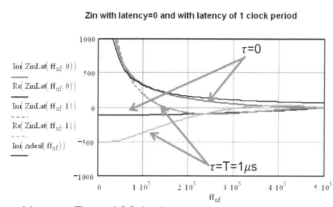

- ## Above $R_{ser}=100$ helps reduce parasitic resistance

Example Negative Cap.: τ=T=1μs, R$_{ser}$=100, C=-5nF, R$_{dac}$=1000

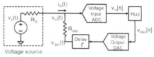

Voltage source

- As next step:
  - Use matthcad to determine Zin for R$_{ser}$:

$Rsa = 50$ $Rsera = 100$ $Ca = -5 \cdot 10^{9}$ $Ta = 1 \cdot 10^{6}$ $Rdaca = 1 \cdot 10^{3}$

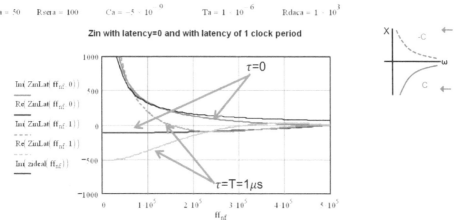

- Above R$_{ser}$=100 helps reduce parasitic resistance

Examples
Positive Capacitance

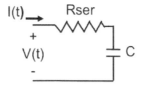

Example Positive Cap.: τ=T=1μs, R_{ser}=100, C=+5nF, R_{dac}=1000

- Same steps as before, then:
  - Enter H(z) into mattlab simuulink to do simulation:

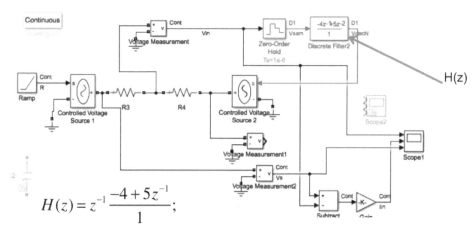

$$H(z) = z^{-1} \frac{-4 + 5z^{-1}}{1};$$

for 1-clock-cycle latency

Example Positive Cap.: τ=T=1μs, R_{ser}=100, C=+5nF, R_{dac}=1000

- As next step:
  - Use mattlab simuulink to do simulation.

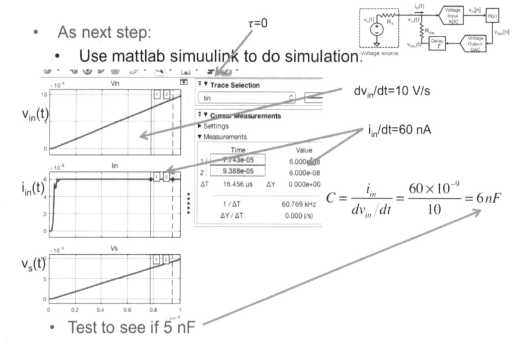

$$C = \frac{i_{in}}{dv_{in}/dt} = \frac{60 \times 10^{-9}}{10} = 6nF$$

- Test to see if 5 nF

Preliminary Data: Positive and Negative Cap

- Preliminary data (alternative topology)

Yellow=Volts

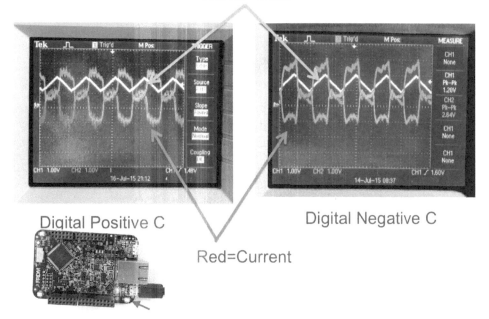

Digital Positive C

Digital Negative C

Red=Current

Phase-Locked Loops

Phase-Locked Loop Design

- Phase-locked loop design
- Used in:
  - Radio receivers
  - Frequency synthesizers
  - Power grid synchronization
- Approach
  - Develop system theory for design of D(z)
  - Second, less complicated design method
    - Start with analog design
    - Convert to digital D(z)

Phase-Locked Loop

- PLL (phase-locked loop) commonly used in radios
- The phase detector can be done with a multiplier
- VCO frequency is proportional to its input voltage
- Also used in power/grid synchronization

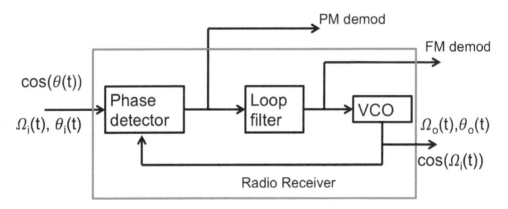

Instantaneous Frequency

- General form

$$g(t) = A \cos(\theta(t))$$

- Define instantaneous frequency

$$\Omega_i(t) = d\, \theta(t) / dt \quad \text{rad/s}$$

- Simple Example:

$$g(t) = A \cos(\Omega_c t + \phi)$$

Then

$$\Omega_i(t) = d\, \theta(t) / dt = d (\Omega_c t + \phi) / dt = \Omega_c$$

Instantaneous Frequency

- Suppose a VCO has the following output:

$$g(t) = A \cos(\Omega_c t) = A \cos((K_V v_f(t)) t)$$

- Plots of instantaneous phase and instantaneous frequency below
- Note: VCO produces a constant output frequency for constant voltage
- VCO produces ramp in phase for constant input voltage – phase integrator

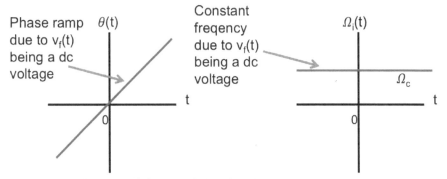

Because $\omega_i(t) = d\, \theta(t) / dt$, variation in phase results in changes in frequency, and changes in frequency result in changes in phase

Phase-Locked Loop Analysis: Multiplier Phase Detector

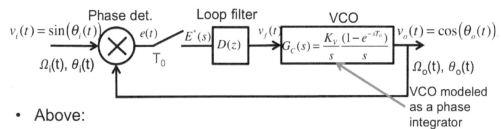

- Above:
  - Use a digital loop filter, D(z) and sampler with T_0 period
  - Use multiplier as phase detector, since:

$$e(t) = A_i \sin\big(\theta_i(t)\big) A_o \sin\big(\theta_o(t)\big) = A_i A_o \sin\big(\Omega_i t + \phi_i\big)\cos\big(\Omega_o t + \phi_o\big)$$

$$= 0.5 A_i A_o \sin\big((\Omega_i - \Omega_o)t + \phi_i - \phi_o\big) + 0.5 A_i A_o \cos\big((\Omega_i + \Omega_o)t + \phi_i + \phi_o\big)$$

$$\approx 0.5 A_i A_o \sin\big((\Omega_i - \Omega_o)t + \phi_i - \phi_o\big) \quad \text{after high frequency is removed}$$

$$= 0.5 A_i A_o \sin\big(\phi_i - \phi_o\big) \quad \text{when frequencies are equal, } \Omega_i = \Omega_o$$

$$\approx 0.5 A_i A_o \big(\phi_i - \phi_o\big) = K_P \big(\phi_i - \phi_o\big) \quad \text{for small angle}$$

- Where phase detector gain coefficient K_P is in volts/radian

Phase-Locked Loop Analysis: Closed-Loop Response

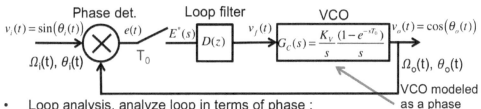

- Loop analysis, analyze loop in terms of phase :
  - First find Gc(z):

$$G_C(s) = K_v \frac{1 - e^{-sT_0}}{s^2} = G'(s)\big(1 - e^{-sT_0}\big); \quad \Rightarrow G'(z) = \frac{K_v z T_0}{(z-1)^2}; \quad \Rightarrow G_C(z) = \frac{K_v z T_0}{(z-1)^2}\frac{(z-1)}{z} = \frac{K_v T_0}{(z-1)}$$

- Find closed-loop pulse-transfer function $G_{CL}(z) =, \theta_o(z)/\theta_i(z)$ as follows:

$$G_{CL}(z) = \frac{\theta_o(z)}{\theta_i(z)} = \frac{D(z)G_C(z)}{1 + D(z)G_C(z)} = \frac{K_p K_v T_0 D(z)}{(z-1) + K_p K_v T_0 D(z)}$$

$$let : D(z) = \frac{s_p}{s + s_p}\bigg|_{s = \frac{2}{T_0}\frac{z-1}{z+1}} = \frac{s_p T_0(z+1)}{2(z-1) + s_p T_0(z+1)} = \frac{s_p T_0(z+1)}{(2 + s_p T_0)z + s_p T_0 - 2}$$

$$so : G_{CL}(z) = \frac{K_p K_v T_0 s_p T_0(z+1)}{z^2(2 + s_p T_0) + z(2 s_p T_0 + s_p T_0 - 4 + K_p K_v T_0 s_p T_0) + (2 - s_p T_0 + K_p K_v T_0 s_p T_0)}$$

- Where VCO coefficient K_V is in radian/s/volt

Phase Locked Loop: Cont. Time Analysis

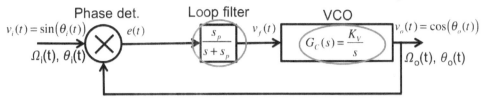

- Discrete-time loop analysis lead to fairly complicated quadratic
- Less complicated but perhaps less precise approach:
  - o Design for natural frequency and damping in s-domain first
  - o Then, use bilinear transform to find D(z) from analog loop filter

$$G_{CL}(s) = \frac{G_C(s)s_p/(s+s_p)}{1+G_C(z)s_p/(s+s_p)} = \frac{K_p K_v s_p}{s(s+s_p)+K_p K_v s_p} = \frac{K_p K_v s_p}{s^2 + s(s_p)+K_p K_v s_p}$$

design for above where, $\omega_n = \sqrt{K_p K_v s_p}$ and damping $\zeta = s_p/(2\omega_n)$

then use bilinear transform for find D(z): $D(z) = \frac{s_p}{s+s_p}\Bigg|_{s=\frac{2}{T_0}\frac{z-1}{z+1}}$

Example PLL Design

Example PLL Step Response from 0.99 to 1 MHz

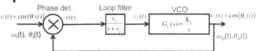

- First design the continuous-time system:
- Then convert loop filter to D(z)

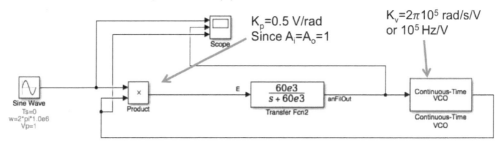

K_p=0.5 V/rad
Since A_i=A_o=1

K_v=$2\pi 10^5$ rad/s/V
or 10^5 Hz/V

so, $\omega_n = \sqrt{K_p K_v s_p} = \sqrt{0.5(2\pi \times 10^5)60,000} = 137,000 \ rad/s, \quad or \ 21.9 \ kHz = 1/46us$

and damping : $\zeta = s_p/(2\omega_n) = 60,000/(274,000)0.22$ which would give approx 50% overshoot

then use bilinear transform for find D(z): $D(z) = \left. \dfrac{s_p}{s+s_p} \right|_{s=\frac{2}{T_0}\frac{z-1}{z+1}}$

Example PLL Step Response from 0.99 to 1 MHz

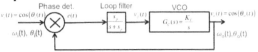

- First design the continuous-time system:
- Then convert loop filter to D(z)

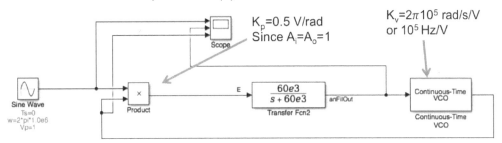

K_p=0.5 V/rad
Since A_i=A_o=1

K_v=$2\pi 10^5$ rad/s/V
or 10^5 Hz/V

so, $\omega_n = \sqrt{K_p K_v s_p} = \sqrt{0.5(2\pi \times 10^5)60,000} = 137,000 \ rad \ / \ s, \quad or \ 21.9 \ kHz = 1/46us$

and damping $:\zeta = s_p /(2\omega_n) = 60,000/(274,000)0.22$ which would give approx 50% overshoot

then use bilinear transform for find D(z): $\quad D(z) = \frac{s_p}{s + s_p}\Bigg|_{s=\frac{2}{T_0}\frac{z-1}{z+1}}$

Example PLL Step Response from 0.99 to 1 MHz

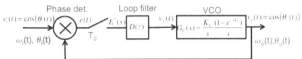

- Discrete-time system:
- Bilinear transform converted loop filter to D(z), using 10 MHz ADC/DAC

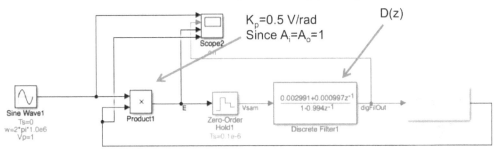

K_p=0.5 V/rad
Since A_i=A_o=1

D(z)

- Note: would not usually have ADC clock > VCO frequency
- But example suffices to illustrate principles within simulator limitations

Example PLL Step Response from 0.99 to 1 MHz

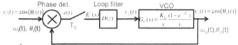

- Compare results continuous-time system:
- Step response 46 us natural freq. period and 50% overshoot expected
- 10 kHz step should result in 0.1 volt change at VCO for 10^5 V/Hz
- Note: would not usually have ADC clock > VCO frequency
- But example suffices to illustrate principles within simulator limitations

Discrete-time system response   Continuous-time system response

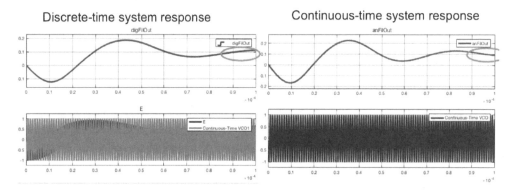

13 APPENDIX

.

Trigonometry Identities

$$\cos(A)\,\cos(B) = 0.5\,\cos(A-B) + 0.5\,\cos(A+B)$$

$$\sin(A)\,\cos(B) = 0.5\,\sin(A-B) + 0.5\,\sin(A+B)$$

$$\sin(A)\sin(B) = 0.5\,\cos(A-B) - 0.5\,\cos(A+B)$$

$$\cos^2(A) = 0.5 + 0.5\cos(2A)$$

$$\sin^2(A) = 0.5 - 0.5\cos(2A)$$

$$a\cos(x) + b\sin(x) = \sqrt{a^2+b^2}\,\cos(x + \arctan(-b/a))$$

$$e^{j\theta} = \cos(\theta) + j\,\sin(\theta)$$

$$e^{j\theta} + e^{-j\theta} = 2\cos(\theta)$$

$$e^{j\theta} - e^{-j\theta} = 2j\sin(\theta)$$

Quadratic formula:

$$ax^2 + bx + c = 0 \quad \Rightarrow \quad x = \frac{-b \pm \sqrt{b^2 - 4ac}}{2a}$$

Laplace Transforms and Z-Transforms

Laplace Trans. $X(s)$	Contin. Time $x(t)$	Sampled Function $x[n]=x(nT_S)$	z-transform. $X(z)$
1	$\delta(t)$	$-$	$-$
$\dfrac{1}{s}; \operatorname{Re}\{s\}>0$	$u(t)$	$u[n]$	$\dfrac{z}{z-1};\ \|z\|>1$
$\dfrac{1}{s^2}; \operatorname{Re}\{s\}>0$	$tu(t)$	$nT_S u[n]$	$\dfrac{zT_S}{(z-1)^2};\ \|z\|>1$
$\dfrac{1}{(s+a)}; \operatorname{Re}\{s\}>-a$	$e^{-at}u(t)$	$e^{-naT_S}u[n]$	$\dfrac{z}{z-e^{-aT_S}};\ \|z\|>\left\|e^{-aT_S}\right\|$
$\dfrac{1}{(s+a)^2}; \operatorname{Re}\{s\}>-a$	$te^{-at}u(t)$	$nT_S e^{-naT_S}u[n]$	$\dfrac{ze^{-aT_S}T_S}{\left(z-e^{-aT_S}\right)^2};\ \|z\|>\left\|e^{-aT_S}\right\|$
$\dfrac{a}{s(s+a)}; \operatorname{Re}\{s\}>0$	$(1-e^{-at})u(t)$	$\left(1-e^{-naT_S}\right)u[n]$	$\dfrac{z\left(1-e^{-aT_S}\right)}{(z-1)\left(z-e^{-aT_S}\right)};\ \|z\|>1$
$\dfrac{s}{s^2+\Omega_0{}^2}; \operatorname{Re}\{s\}>0$	$\cos(\Omega_0 t)u(t)$	$\cos(n\Omega_0 T_S)u[n]$	$\dfrac{z^2-z\cos(\Omega_0 T_S)}{z^2-2z\cos(\Omega_0 T_S)+1};\ \|z\|>1$
$\dfrac{\Omega_0}{s^2+\Omega_0{}^2}; \operatorname{Re}\{s\}>0$	$\sin(\Omega_0 t)u(t)$	$\sin(n\Omega_0 T_S)u[n]$	$\dfrac{z\sin(\Omega_0 T_S)}{z^2-2z\cos(\Omega_0 T_S)+1};\ \|z\|>1$

Z-transform Pairs

Discrete-time Function	z-transform
$\delta[n]$	1
$u[n]$	$\dfrac{z}{z-1};\ \ |z|>1$
$nu[n]$	$\dfrac{z}{(z-1)^2};\ \ |z|>1$
$a^n u[n]$	$\dfrac{z}{z-a};\ \ |z|>|a|$
$-a^n u[-n-1]$	$\dfrac{z}{z-a};\ \ |z|<|a|$
$na^n u[n]$	$\dfrac{az}{(z-a)^2};\ \ |z|>|a|$
$\cos(\omega_0 n)u[n]$	$\dfrac{z^2 - z\cos(\omega_0)}{z^2 - 2z\cos(\omega_0)+1};\ \ |z|>1$
$\sin(\omega_0 n)u[n]$	$\dfrac{z\sin(\omega_0)}{z^2 - 2z\cos(\omega_0)+1};\ \ |z|>1$

Modified Z-Transforms

Mod. z-transform, X(z,m)	z-transform	Time Func	Laplace Transform
	1	$\delta(t)$	1
$\dfrac{1}{z-1}$	$\dfrac{z}{z-1};\ |z|>0$	$u(t)$	$1/s$
$\dfrac{mT_0}{z-1}+\dfrac{T_0}{(z-1)^2}$	$\dfrac{zT_0}{(z-1)^2};\ |z|>1$	$tu(t)$	$1/s^2$
$\dfrac{e^{-amT_0}}{z-e^{-aT_0}}$	$\dfrac{z}{z-e^{-aT_0}};\ |z|>e^{-aT_0}$	$e^{-at}u(t)$	$\dfrac{1}{s+a}$
$\dfrac{T_0 e^{-amT_0}\left[e^{-aT_0}+m\left(z-e^{-aT_0}\right)\right]}{\left(z-e^{-aT_0}\right)^2}$	$\dfrac{T_0 z e^{-aT_0}}{\left(z-e^{-aT_0}\right)^2};\ |z|>e^{-aT_0}$	$te^{-at}u(t)$	$\dfrac{1}{(s+a)^2}$

Fourier Transform in "f"

Fourier Transform Pairs $\quad X(f) = \int_{-\infty}^{\infty} x(t)e^{-j2\pi ft}dt \quad\quad x(t) = \int_{-\infty}^{\infty} X(f)e^{j2\pi ft}df$

$\delta(t) \leftrightarrow 1$	$\cos(2\pi f_0 t) \leftrightarrow 0.5(\delta(f+f_0) + \delta(f-f_0))$
$1 \leftrightarrow \delta(f)$	$\sin(2\pi f_0 t) \leftrightarrow 0.5j(\delta(f+f_0) - \delta(f-f_0))$
$u(t) \leftrightarrow \dfrac{1}{2}\delta(f) + \dfrac{1}{j2\pi f}$	$sgn(t) \leftrightarrow \dfrac{1}{j\pi f}$
$\Pi(t/\tau) \leftrightarrow \tau \cdot sinc(\pi f\tau)$	$2B\, sinc(2\pi Bt) \leftrightarrow \Pi(f/(2B))$
$\Delta\left(\dfrac{t}{\tau}\right) \leftrightarrow \dfrac{\tau}{2} \cdot sinc^2(\pi f\tau/2)$	$B\, sinc^2(\pi Bt) \leftrightarrow \Delta(f/(2B))$
$e^{j2\pi f_0 t} \leftrightarrow \delta(f-f_0)$	$e^{-at}u(t) \leftrightarrow \dfrac{1}{a+j2\pi f}$
$e^{-t^2/(2\sigma^2)} \leftrightarrow \sigma\sqrt{2\pi}\, e^{-2(\sigma\pi f)^2}$	$e^{-a|t|} \leftrightarrow \dfrac{2a}{a^2 + (2\pi f)^2}$

Fourier Transform Properties

$g(t)e^{j2\pi f_0 t} \leftrightarrow G(f-f_0)$	$g(t-t_0) \leftrightarrow G(f)e^{-j2\pi t_0 f}$
$g(at) \leftrightarrow \dfrac{1}{\|a\|}G\left(\dfrac{f}{a}\right)$	$G(t) \leftrightarrow g(-f)$
$g(t) * h(t) \leftrightarrow G(f)H(f)$	$g(t)h(t) \leftrightarrow G(f) * H(f)$
$\dfrac{dg(t)}{dt} \leftrightarrow j2\pi f\, G(f)$	$\displaystyle\int_{-\infty}^{t} g(\alpha)d\alpha \leftrightarrow \dfrac{G(f)}{j2\pi f} + G(0)\delta(f)/2$
	$\displaystyle\int_{-\infty}^{\infty} |g(t)|^2 dt = \int_{-\infty}^{\infty} |G(f)|^2 df$

Fourier Transform in "Ω"

Fourier Transform Pairs $X(\Omega) = \int_{-\infty}^{\infty} x(t)e^{-j\Omega t}dt$ $x(t) = \frac{1}{2\pi}\int_{-\infty}^{\infty} X(\Omega)e^{j\Omega t}d\Omega$

$\delta(t) \leftrightarrow 1$	$\cos(\Omega_0 t) \leftrightarrow \pi(\delta(\Omega + \Omega_0) + \delta(\Omega - \Omega_0))$
$1 \leftrightarrow 2\pi\delta(\Omega)$	$\sin(\Omega_0 t) \leftrightarrow j\pi(\delta(\Omega + \Omega_0) - \delta(\Omega - \Omega_0))$
$u(t) \leftrightarrow \pi\delta(\Omega) + \dfrac{1}{j\Omega}$	$sgn(t) \leftrightarrow \dfrac{2}{j\Omega}$
$\Pi(t/\tau) \leftrightarrow \tau \cdot sinc(\Omega\tau/2)$	$W\,sinc(Wt) \leftrightarrow \pi\,\Pi(\Omega/(2W))$
$\Delta\left(\dfrac{t}{\tau}\right) \leftrightarrow \dfrac{\tau}{2} \cdot sinc^2(\Omega\tau/4)$	$W\,sinc^2(Wt/2) \leftrightarrow 2\pi\Delta(\Omega/(2W))$
$e^{j\Omega_0 t} \leftrightarrow 2\pi\delta(\Omega - \Omega_0)$	$e^{-at}u(t) \leftrightarrow \dfrac{1}{a + j\Omega}$
$e^{-t^2/(2\sigma^2)} \leftrightarrow \sigma\sqrt{2\pi}\,e^{-\sigma^2\Omega^2/2}$	$e^{-a|t|} \leftrightarrow \dfrac{2a}{a^2 + \Omega^2}$

Fourier Transform Properties

$g(t)e^{j\Omega_0 t} \leftrightarrow G(\Omega - \Omega_0)$	$g(t - t_0) \leftrightarrow G(\Omega)e^{-jt_0\Omega}$
$g(at) \leftrightarrow \dfrac{1}{|a|}G\left(\dfrac{\Omega}{a}\right)$	$G(t) \leftrightarrow 2\pi g(-\Omega)$
$g(t) * h(t) \leftrightarrow G(\Omega)H(\Omega)$	$g(t)h(t) \leftrightarrow \dfrac{1}{2\pi}G(\Omega) * H(\Omega)$
$\dfrac{dg(t)}{dt} \leftrightarrow j\Omega\,G(\Omega)$	$\displaystyle\int_{-\infty}^{t} g(\alpha)d\alpha \leftrightarrow \dfrac{G(\Omega)}{j\Omega} + \pi G(0)\delta(\Omega)$
	$\displaystyle\int_{-\infty}^{\infty} |g(t)|^2 dt = \dfrac{1}{2\pi}\int_{-\infty}^{\infty} |G(\Omega)|^2 d\Omega$

Q Table

Table of Q(α)

α	Q(α)
0	0.5000
0.5	0.3085
1.0	0.1587
1.5	0.0668
2.0	0.0228
2.5	0.0062
3.0	0.0014

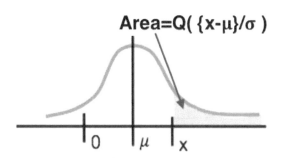

Area=Q({x-μ}/σ)

Matrix Identities

$$\left(\overline{\overline{A}}\, \overline{\overline{B}} \right)^{-1} = \overline{\overline{B}}^{-1}\, \overline{\overline{A}}^{-1}$$

$$\left(\overline{\overline{A}}\, \overline{\overline{B}} \right)^{H} = \overline{\overline{B}}^{H}\, \overline{\overline{A}}^{H}$$

$$\left(\overline{\overline{A}} + \overline{\overline{B}} \right)^{H} = \overline{\overline{A}}^{H} + \overline{\overline{B}}^{H}$$

$$\left(\overline{\overline{A}}\, \overline{\overline{B}} \right)^{T} = \overline{\overline{B}}^{T}\, \overline{\overline{A}}^{T}$$

$$\left(\overline{\overline{A}}\, \overline{\overline{B}} \right)^{T} = \overline{\overline{B}}^{T}\, \overline{\overline{A}}^{T}$$

$$\left(\overline{\overline{A}}^{H} \right)^{-1} = \left(A^{-1} \right)^{H}$$

$$\left(A^{T} \right)^{-1} = \left(A^{-1} \right)^{T}$$

Inverse Laplace Transform

Laplace Inverse Transform in Depth

$$x(t) = L^{-1}\{X(s)\} = \frac{1}{j2\pi} \int_{\sigma-\infty}^{\sigma+\infty} X(s)e^{st}\,ds$$

- For both 1-sided and 2-sided Laplace, inverse Laplace transform is more difficult because:
  - ○ "s" and X(s) are 2-dimensional (real, imaginary)
  - ○ Compared to 1-D "t" variable in forward transform
- Will investigate/analyze inverse Laplace using vector fields, curl, and Stokes' theorem
  - ○ To more deeply understand inverse Laplace transform
  - ○ To make "use of" and make "connections between" math tools already at an engineering student's disposal

$$\nabla \times \vec{E} = \begin{vmatrix} \hat{x} & \hat{y} & \hat{z} \\ \partial/\partial x & \partial/\partial y & \partial/\partial z \\ E_x & E_y & E_z \end{vmatrix} \qquad \oint_{line} \vec{E}\cdot d\vec{l} = \int_{area} (\nabla \times \vec{E})\cdot d\vec{a}$$

Mappings and Vector Fields

- Mapping of Laplace: $(g(t) \in \mathcal{C}, t \in \mathcal{R}) \rightarrow (G(s) \in \mathcal{C}, s \in \mathcal{C})$
  where $\mathcal{R}$ is set of real numbers, $\mathcal{C}$ is set of complex numbers
  - ○ or $\mathcal{L}: (\mathcal{C}, \mathcal{R}) \rightarrow (\mathcal{C}, \mathcal{C})$ ...or $\mathcal{L}: \mathcal{R}^3 \rightarrow \mathcal{R}^4$, maps 3-D into 4-D
  - ○ Note: function g(t) maps $g: \mathcal{R} \rightarrow \mathcal{C}$, and G(s) maps $G: \mathcal{C} \rightarrow \mathcal{C}$
- Laplace transform can be represented as vector field

$$\vec{G}(s) = \begin{bmatrix} G_r(s) \\ G_i(s) \end{bmatrix} = \begin{bmatrix} G_r\left(\begin{bmatrix} \sigma \\ \Omega \end{bmatrix}\right) \\ G_i\left(\begin{bmatrix} \sigma \\ \Omega \end{bmatrix}\right) \end{bmatrix}$$

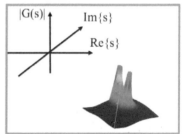

$$\vec{G_3}(s) = \begin{bmatrix} G_r(s) \\ G_r(s) \\ 0 \end{bmatrix}$$

- Or even extended to a 3-D vector where now $\mathcal{L}: (\mathcal{C}, \mathcal{R}) \rightarrow (G(s) \in \mathcal{R}^3, s \in \mathcal{C})$
- This 3D representation is of use later

Complex Analysis: Analytic Functions

- Complex function G(s), G: $\mathcal{C} \to \mathcal{C}$ is analytic if it is "well behaved"
  - o Has finite well-defined derivative at all points in the region
    - – Implies single-valued, continuous, has Taylor series
    - – *Often constrained to some region* of s-plane
- Examples of analytic functions:
  - o s, e^{st}, s^2, s^{-2} for s≠0, N(s)/D(s) *except at poles*, ln(s) for s≠0
- Example of not analytic: G(s)=s* (derivative changes with direction)

$$\frac{dG(s)}{ds} = \frac{ds^*}{ds} \Rightarrow \lim_{\Delta s \to 0} \frac{G(s+\Delta s) - G(s)}{\Delta s} = \lim_{\Delta s \to 0} \frac{(s+\Delta s)^* - s^*}{\Delta s}$$

$$\sigma\text{-axis}: \quad = \lim_{\Delta \sigma \to 0} \frac{(s+\Delta \sigma)^* - s^*}{\Delta \sigma} = \frac{\Delta \sigma}{\Delta \sigma} = 1$$

Derivative depends on direction, ouch!
NOT well-defined

$$\Omega\text{-axis}: \quad = \lim_{\Delta \Omega \to 0} \frac{(s+j\Delta \Omega)^* - s^*}{j\Delta \Omega} = \frac{-j\Delta \Omega}{j\Delta \Omega} = -1$$

Complex Analysis: Cauchy-Riemann Equations

- Example of analytic: G(s)=s

$$\frac{dG(s)}{ds} = \frac{ds}{ds} \Rightarrow \lim_{\Delta s \to 0} \frac{G(s+\Delta s) - G(s)}{\Delta s} = \lim_{\Delta s \to 0} \frac{(s+\Delta s) - s}{\Delta s}$$

$$\sigma\text{-axis}: \quad = \lim_{\Delta \sigma \to 0} \frac{(s+\Delta \sigma) - s}{\Delta \sigma} = \frac{\Delta \sigma}{\Delta \sigma} = 1$$

Derivative does not depends on direction.
Yes, well-defined.

$$\Omega\text{-axis}: \quad = \lim_{\Delta \Omega \to 0} \frac{(s+j\Delta \Omega) - s}{j\Delta \Omega} = \frac{j\Delta \Omega}{j\Delta \Omega} = 1$$

- Generalizing: require $d/d\sigma = d/dj\Omega$ for G(s) to be analytic:

$$require: \frac{\partial G(s)}{\partial \sigma} = \frac{\partial G(s)}{j\partial \Omega} \Rightarrow \frac{\partial (G_r(s) + jG_i(s))}{\partial \sigma} = \frac{\partial (G_r(s) + jG_i(s))}{j\partial \Omega}$$

equating real and imaginary parts:

$$\frac{\partial G_r(s)}{\partial \sigma} = \frac{\partial G_i(s)}{\partial \Omega} \quad and \quad \frac{\partial G_i(s)}{\partial \sigma} = \frac{-\partial G_r(s)}{\partial \Omega}$$

Necessary conditions for analytic function.

- These are Cauchy-Riemann equations/conditions
- Riemann is student of Gauss~1851 g[11]

Complex Analysis: Path Independent Integral

- Every analytic function has path-independent contour integrals
  - o Important because inverse Laplace is a contour integral
  - o So, one is free to choose the contour in analytic region
- Consider an arbitrary G(s) integration over contour C which may be open (not closed), and rearrange it to a dot product form:

Note: this is not a dot product here!

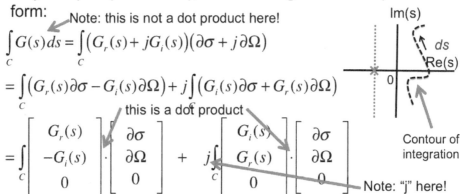

$$\int_C G(s)\,ds = \int_C (G_r(s) + jG_i(s))(\partial\sigma + j\partial\Omega)$$

$$= \int_C (G_r(s)\partial\sigma - G_i(s)\partial\Omega) + j\int_C (G_i(s)\partial\sigma + G_r(s)\partial\Omega)$$

this is a dot product

$$= \int_C \begin{bmatrix} G_r(s) \\ -G_i(s) \\ 0 \end{bmatrix} \cdot \begin{bmatrix} \partial\sigma \\ \partial\Omega \\ 0 \end{bmatrix} + j\int_C \begin{bmatrix} G_i(s) \\ G_r(s) \\ 0 \end{bmatrix} \cdot \begin{bmatrix} \partial\sigma \\ \partial\Omega \\ 0 \end{bmatrix}$$

Contour of integration

Note: "j" here!

- Final step, we must show these are conservative

Complex Analysis: Conservative Fields

- A conservative field
  - Has zero curl, which implies path-independent line integrals
- This true because of Stokes' theorem:

$$\oint_{line} \vec{E} \cdot d\vec{l} = \int_{area} (\nabla \times \vec{E}) \cdot d\vec{a} = 0 \quad \text{if} \quad \nabla \times \vec{E} = 0$$

Area implies no "holes" in region

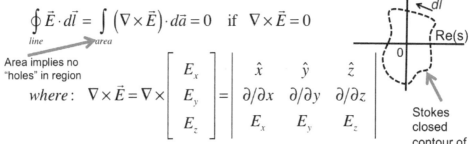

$$where: \quad \nabla \times \vec{E} = \nabla \times \begin{bmatrix} E_x \\ E_y \\ E_z \end{bmatrix} = \begin{vmatrix} \hat{x} & \hat{y} & \hat{z} \\ \partial/\partial x & \partial/\partial y & \partial/\partial z \\ E_x & E_y & E_z \end{vmatrix}$$

Stokes closed contour of integration

$$and: \quad \vec{A} \cdot \vec{B} = A^T B = A_x E_x + A_y E_y + A_z E_z = |AB|\cos(\theta)$$

- Example: electro_static_ field is conservative, because dB/dt=0:

$$\nabla \times \vec{E} = -d\vec{B}/dt = 0 \quad \text{if} \quad d\vec{B}/dt = 0$$

$$so: \quad \oint_{line} \vec{E} \cdot d\vec{l} = \int_{area} (\nabla \times \vec{E}) \cdot d\vec{a} = 0 \quad \text{and is conservative!}$$

Complex Analysis: Why Path-independent

- A conservative field
  - Has zero curl, which implies path-independent line integrals
- This true because of Stokes' theorem
- Both solid red contours must have same integral, because when added to the dashed blue contour (of whatever fixed value), the net result is a closed contour with value=0.
- So, the two sold red contours have equal integrals
- Notion is similar to work moving a charge in electrostatics

Im(s)

Dashed blue open contour of integration in dashed blue ——→

dl

Re(s)

0

Various solid red contours that form closed contour when combined with dashed blue

Complex Analysis: Path Independent Integral

- Finally, show that every analytic function has path-independent contour integrals
- So, **closed contour integral=0,** *if no poles inside contour*

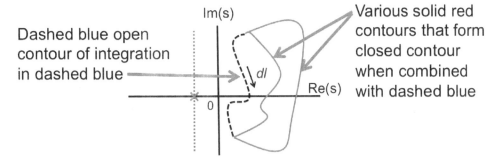

$$\oint_c G(s)ds = 0 = \int_c \begin{bmatrix} G_r(s) \\ -G_i(s) \\ 0 \end{bmatrix} \cdot \begin{bmatrix} \partial\sigma \\ \partial\Omega \\ 0 \end{bmatrix} + j\int_c \begin{bmatrix} G_i(s) \\ G_r(s) \\ 0 \end{bmatrix} \cdot \begin{bmatrix} \partial\sigma \\ \partial\Omega \\ 0 \end{bmatrix}$$

Cauchy-Goursat Theorem
$$\oint_c G(s)ds = 0$$

because the 2 field terms are conservative, both with curl=0:

$$\begin{vmatrix} \hat\sigma & \hat\Omega & \hat z \\ \partial/\partial\sigma & \partial/\partial\Omega & \partial/\partial z \\ G_r(s) & -G_i(s) & 0 \end{vmatrix} = \left(-\frac{\partial G_i(s)}{\partial\sigma} - \frac{\partial G_r(s)}{\partial\Omega} \right)\hat z = 0 \quad \text{since} \quad \frac{\partial G_i(s)}{\partial\sigma} = \frac{-\partial G_r(s)}{\partial\Omega}$$

$$\begin{vmatrix} \hat\sigma & \hat\Omega & \hat z \\ \partial/\partial\sigma & \partial/\partial\Omega & \partial/\partial z \\ G_i(s) & G_r(s) & 0 \end{vmatrix} = \left(\frac{\partial G_r(s)}{\partial\sigma} - \frac{\partial G_i(s)}{\partial\Omega} \right)\hat z = 0 \quad \text{since} \quad \frac{\partial G_r(s)}{\partial\sigma} = \frac{\partial G_i(s)}{\partial\Omega}$$

- Curls are zero, because of Cauchy-Riemann conditions
  But remember from Stokes' theorem: no holes in region

Cauchy Integral Theorems

1. Closed contour includes no poles
   - See previous slide

$$\oint_C G(s)\,ds = 0$$

From Stokes, no holes in region

2. Closed contour includes multiple pole or zero
   - $G(s) = (s+a)^n$ for $n \neq 1$

$$\oint_C (s+a)^n\,ds = \left.\frac{(s+a)^{n+1}}{n+1}\right|_{S_1}^{S_1} = 0 \quad for \quad n \neq 1$$

Closed contour
$s1 \rightarrow s1 \implies \int = 0$

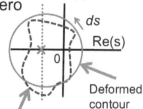

3. Closed contour includes simple pole
   - $G(s) = (s+a)^{-1}$

$$\oint_C \frac{1}{(s+a)}\,ds = \int_{-\pi}^{\pi} \frac{1}{(re^{j\theta} - a + a)}\,jre^{j\theta}\,d\theta$$

$$= \int_{-\pi}^{\pi} \frac{jre^{j\theta}}{re^{j\theta}}\,d\theta = \int_{-\pi}^{\pi} j\,d\theta = jd\theta\Big|_{-\pi}^{\pi} = j2\pi$$

Key result

where: $ds/d\theta = d(re^{j\theta} - a)/d\theta = jre^{j\theta} \implies ds = jre^{j\theta}\,d\theta$

Original blue-dashed contour of integration

Here we change contour of integration to red circle of radius r, centered around "a," because path-independent!

Complex Analysis: Cauchy Residue Theorem

- Suppose an analytic function G(s) in a region
- Then, add a pole by multiplying G(s) by 1/(s+a)
- Then, use partial fraction expansion to remove the pole
- Leaves a remainder that is analytic without a pole at "a"
- So:

Remember from Stokes, no holes in region

$$\oint_C \frac{G(s)}{(s+a)}\,ds = \oint_C \left(\frac{G(s)|_{s=-a}}{(s+a)} + remainder(s) \right)ds = j2\pi G(s)|_{s=-a}$$

Also implied: contour does not pass through the pole at "a"
Remaining terms after partial fraction expansion, removes s+a term,
results in analytic remainder with no poles at "a" and integral=0

- Generalizing to full partial fraction expansion

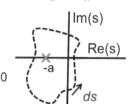

Cauchy Residue theorem (assuming first-order poles)

$$\oint_C \frac{G(s)}{\prod_n (s+a_n)}\,ds = \oint_C \left(\sum_n \frac{residue(a_n)}{s+a_n} \right)ds = j2\pi \sum_n residues(a_n)$$

Complex Analysis: Inverse Laplace of 1/(s+a)

- Now, compare Cauchy integral theorem with Inverse Laplace transform of $G(s)=1/(s+a)$

Remember from Stokes, no holes in region

Cauchy integral theorem: $\displaystyle\oint_C \frac{G(s)}{(s+a)}ds = j2\pi G(-a)$

Closed contour

Inverse Laplace transform:

$$L^{-1}\left\{\frac{1}{s+a}\right\} = \frac{1}{j2\pi}\int_{\sigma_i-j\infty}^{\sigma_i+j\infty}\frac{1}{s+a}e^{st}\,ds = \frac{1}{j2\pi}\int_{\sigma_i-j\infty}^{\sigma_i+j\infty}\frac{e^{st}}{s+a}\,ds$$

Note:

Open contour

$$L\{dx/dt + ax = 0\} = sX(s)+aX(s)=0 \quad\Rightarrow\quad X(s)=1/(s+a)$$

- Cauchy requires closed contour
- But Laplace is open contour, e^{st} replaces G(s)
- So, see next slide

Complex Analysis: Inverse Laplace of 1/(s+a)

- So, divide contour into C1 and C2 to use Cauchy integral
- Suppose contour below, as $r\to\infty$

$$\oint_C \frac{G(s)}{(s+a)}ds = j2\pi G(-a) = \lim_{r\to\infty}\left(\int_{C1}\frac{G(s)}{(s+a)}ds + \int_{C2}\frac{G(s)}{(s+a)}ds\right)$$

Red-dashed contour C2 closing contour

$$\oint_C \frac{e^{st}}{s+a}ds = \lim_{r\to\infty}\left(\int_{\sigma_i-rj\sin(\phi)}^{\sigma_i+rj\sin(\phi)}\frac{e^{st}}{s+a}ds + \int_{-\arccos(\sigma_i/r)}^{\arccos(\sigma_i/r)}\frac{e^{(re^{j\phi})t}}{re^{j\phi}+a}jre^{j\phi}\,d\phi\right)$$

Im(s)

$$= \int_{\sigma_i-j\infty}^{\sigma_i+j\infty}\frac{e^{st}}{s+a}ds = j2\pi G(-a) = j2\pi e^{-at}u(t)$$

$where: s = re^{j\phi}$ on C2 $\Rightarrow$ $ds = jre^{j\phi}d\phi$ on C2

ds   r   ds ϕ Re(s)   -a   0 σ_i

and suffice to say that it can be shown the integral on C2 $\to 0$:

thus Laplace : $\displaystyle\frac{1}{j2\pi}\int_{\sigma_i-j\infty}^{\sigma_i+j\infty}\frac{e^{st}}{s+a}ds = e^{-at}u(t)$

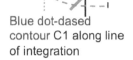

- Where u(t) was added since we are using 1-sided Laplace transform

Blue dot-dashed contour C1 along line of integration

Complex Analysis & Laplace Summary

- Differential equations underlie most systems
- est is an eigenfunction of differential equations (derivatives)
- Laplace transform used to solve differential equations
- Analytic functions: well-defied derivative in s-plane: est, s^2, s^{-2} except s=0
- Cauchy-Riemann Equations: satisfied by analytic functions
- Stokes' theorem shows conservative field has path-independent integral
- Conservative field has zero curl
- Analytic function has zero-curl in its vector-field form
- Analytic functions have path-independent integral
- Closed contour integral in analytic region without poles or holes =0
- Closed contour integral in analytic region with multiple poles or zeroes =0
- Closed contour integral in analytic region with single pole = j2π
- Cauchy Residue Theorem forms basis for inverse Laplace:

$$\oint_C \frac{G(s)}{s+a}ds = \oint_C \left(\frac{G(-a)}{s+a} + \text{remainder}(s)\right)ds = j2\pi G(-a) \quad \Rightarrow \quad x(t) = \frac{1}{j2\pi}\int_{\sigma_i-j\infty}^{\sigma_i+j\infty} X(s)ds$$

Also: Analytic Functions and Laplace's Equation

- Another fact about analytic functions:
- Analytic functions satisfy Laplace's equation
- Shown by substituting one Cauchy-Riemann into the other:

$$require: \frac{\partial G(s)}{\partial \sigma} = \frac{\partial G(s)}{j\partial\Omega} \Rightarrow \frac{\partial(G_r(s)+jG_i(s))}{\partial\sigma} = \frac{\partial(G_r(s)+jG_i(s))}{j\partial\Omega}$$

recall the Cauchy-Riemann equations:

$$\frac{\partial G_r(s)}{\partial\sigma} = \frac{\partial G_i(s)}{\partial\Omega} \quad and \quad \frac{\partial G_i(s)}{\partial\sigma} = \frac{-\partial G_r(s)}{\partial\Omega}$$

taking derivatives of both equations:

$$\frac{\partial^2 G_r(s)}{\partial\sigma^2} = \frac{\partial^2 G_i(s)}{\partial\sigma\,\partial\Omega} \quad and \quad \frac{\partial^2 G_i(s)}{\partial\sigma\,\partial\Omega} = \frac{-\partial^2 G_r(s)}{\partial\Omega^2}$$

substituting one into the other yields:

$$\frac{\partial^2 G_r(s)}{\partial\sigma^2} + \frac{\partial^2 G_r(s)}{\partial\Omega^2} = 0 \quad or \quad \nabla^2 G_r(s) = \nabla\cdot\nabla G_r(s) = 0$$

Recall the definition of divergence:
$$\nabla\cdot\vec{F}(x,y,z) = \begin{bmatrix}\partial/\partial x\\\partial/\partial y\\\partial/\partial z\end{bmatrix}\cdot\begin{bmatrix}F_x(x,y,z)\\F_y(x,y,z)\\F_z(x,y,z)\end{bmatrix}$$
$$= \partial F_x/\partial x + \partial F_y/\partial y + \partial F_z/\partial z$$

Recall the definition of gradient:
$$\nabla\cdot v(x,y,z) = \begin{bmatrix}\partial v/\partial x\\\partial v/\partial y\\\partial v/\partial z\end{bmatrix}$$

- By analogy, recall for scalar electro*static* voltage potential v

$$\nabla^2 v(x,y,z) = \frac{\partial^2 v}{\partial x^2} + \frac{\partial^2 v}{\partial y^2} + \frac{\partial^2 v}{\partial z^2} = 0 \quad and \quad E = -\nabla v(x,y,z) \text{ for electrostatics}$$

Recall: work to move charge in electrostatic field

- Fields such as E= -∇v(x,y,z) are conservative (curl=0) since:
  o Vector identity ∇×∇·(∇V)= ∇×∇²V =0 for any vector V

END

Made in United States
Troutdale, OR
01/03/2025

27575147R00137